Short Fibre
Reinforced
Thermoplastics

POLYMER ENGINEERING RESEARCH STUDIES SERIES

Series Editor: **Professor M. J. Bevis**
School of Materials, Brunel University, Uxbridge, England

1. Short Fibre Reinforced Thermoplastics
   M. J. Folkes

Short Fibre Reinforced Thermoplastics

M. J. Folkes

Brunel University, Uxbridge, England

RESEARCH STUDIES PRESS
A DIVISION OF JOHN WILEY & SONS LTD.
Chichester · New York · Brisbane · Toronto · Singapore

RESEARCH STUDIES PRESS

Editorial Office:
8 Willian Way, Letchworth, Herts. SG6 2HG, England

Copyright © 1982 by John Wiley & Sons Ltd.

British Library Cataloguing in Publication Data:

Folkes, M. J.
  Short fibre reinforced thermoplastics.—(Polymer engineering research studies)
  1. Thermoplastics
  I. Title   II. Series
  668.4′23   TP1180.T5

  ISBN 0 471 10209 1

Printed in Great Britain

Preface

When thermoplastics containing short fibres were first introduced
onto the market it was with the intention of producing a range of
new materials possessing properties that were intermediate between
the high tonnage commodity plastics and the sophisticated continuous
fibre reinforced composites, well established in the aerospace
industry. The increase in stiffness and strength of the short fibre
composite compared to the parent thermoplastic was modest but
nevertheless sufficient to enable this class of material to penetrate
into lightly stressed engineering applications. However during the
last few years there have been some significant advances. We have
seen the emergence of the very high melting point thermoplastics
together with a gradual reduction in the costs of the specialist
fibres such as carbon. Material manufacturers are now combining
engineering thermoplastics with these more expensive fibres to produce
a new range of products having properties that are approaching those
of the traditional long fibre composites. There is still a long way
to go, however, and significant improvements in materials design and
fabrication technology are needed in order to optimize these
developing reinforced thermoplastics.

This book has been written primarily for the plastics industry and
describes some of the concepts on which short fibre reinforcement
are based and which can be used to develop products having specified
properties. Since it is the intention that this series of books
concerned with Polymer Engineering should be fairly concise, it is
impossible to give both depth and breath to the subject, in the
space available. As far as the present book is concerned, it was
decided to restrict the discussion to mechanical properties and not
to give details of commercially available materials and processing

equipment. The latter information, if given, would have been rapidly out of date and is best sought directly from the appropriate manufacturers. The result is an attempt to concentrate on the principles of short fibre reinforced thermoplastics. To this end, only a minimum of mathematics has been used – enough to develop the subject quantitatively but not sufficient, I hope, to deter the reader from exploring the book beyond this preface.

M. J. Folkes

Acknowledgements

I would like to thank Professor M. Bevis for encouraging me to write
this book and for his critical appraisal of my draft manuscript.
A number of members of my research group kindly provided photographs
resulting from their own projects - Dr D. Kells, figures 2.1 and
5.21; Mr R.H. Burton, figures 3.14, 5.4 and 5.11; Mr J. Sharp,
figure 5.8. I am grateful to Dr S. Turner, ICI Petrochemicals and
Plastics Division for the provision of figures 7.15 and 7.17.

I would like to thank all the authors and publishers for granting
me permission to reproduce data and diagrams from journal articles,
in particular:-

The Institute of Physics: Fig. 2.2
Applied Science Publishers Ltd: Figs. 7.12 and 7.13
John Wiley and Sons Inc.: Figs. 3.13, 4.4 and 4.5
Marcel Dekker Inc. : Figs. 3.7 and 7.14
Society of Plastics Engineers Inc.: Figs. 3.1, 3.2, 3.4-3.6, 3.8,3.11,
                                    4.1, 6.1-6.8, 6.12 and 7.18
Chapman and Hall Ltd.: Figs. 5.3, 5.5, 5.7, 5.10, 5.15-5.17, 5.22 and
                       5.23
IPC Science and Technology Press Ltd.: Figs. 2.7, 3.9, 3.10, 3.12,
                                       4.3, 5.2, 5.14, 6.11 and 7.4
Plastics and Rubber Institute.: Figs. 5.1, 5.9, 5.12, 5.13, 6.13-6.16,
                                7.1, 7.5-7.8 and 7.16

Copyright of these individual publishers is acknowledged. Figs. 7.2,
7.3, 7.9-7.11 are reproduced with the permission of the Head of
Information Services, National Physical Laboratory, Teddington.
Fig. 5.14 is also reproduced with the permission of the Controller
of Her Majesty's Stationery Office.

Last but not least, I would like to thank Miss L. Rolph for the
patience and care that she has shown in the typing of the manuscript
in camera ready form and to Mrs R. Pratt and Mr K. Batchelor for
the tedious business of preparing the photographs.

Contents

CHAPTER 1
Introduction

While Leo Baekeland recognized that wood flour was an essential
additive in phenolic moulding powder formulations, this and
other fillers were not considered then as essential ingredients for
thermoplastics. As long as thermoplastics were inexpensive and
plentiful, there was little enthusiasm for the idea of producing
filled grades, at least before the OPEC price increases. However,
such a need became apparent in the mid 1970's and so it is not
surprising to read that about 1 million tonnes of filler were used
by the American plastics industry in 1980 with more than 10% of
this being used in thermoplastics. The range of fillers available is
very large, extending from the mineral and inorganic fillers to the
more expensive fibres such as glass, carbon and Kevlar. In principle
there is really no limitation on the number of possible permutations
of matrix and filler that can be combined to produce a composite.
However, only a limited number have remained after the initial and
extensive research and development efforts by many companies.
Commercially available filled thermoplastics appear to fall into two
broad categories. There are those that are based on comparatively
cheap and abundant particulate fillers and ideally whose function
is to reduce the overall costs of the composite, and yet at the
same time to slightly improve the load bearing capabilities of the
material or to impart special properties to the composite. Then
there are those thermoplastics containing short fibres whose
function is much more than just a simple filler. As Kelly (1973)
has pointed out, fibre reinforced materials are developed in order

to exploit the properties of the stiff and strong fibres and the
plastic is used because it is a suitable "binder" and can be easily
moulded. This is a particularly important point in the context of
this book, since the mechanical properties of many short fibre
reinforced thermoplastics fall well short of their maximum realisable
values. While this has not been a serious drawback in the past, it
is certain that major improvements in the fabrication and utilization
of fibre reinforced thermoplastics will be needed. Many more
reinforced thermoplastics are being used in critical load bearing
applications, some of which also demand minimum component weight.
The need for high specific stiffness and strength coupled with the
ability to operate continuously at elevated temperatures requires
that the reinforcing effect of the fibres is used to maximum
advantage. These requirements have to be met by an appropriate
understanding of the effects of processing conditions on the
microstructure and properties of the final component. The need for
improved control of fibre length and orientation during the
compounding and moulding stages is particularly vital when expensive
combinations of fibre and matrix are being considered. A 13% annual
growth for speciality thermoplastics e.g. polyethersulphones,
polyetherketones, polyimides etc is forecast for the period until
1985. At the same time, speciality fibres such as carbon and Kevlar
are expected to have an annual growth of about 9%. Carbon fibre
reinforced polyetheretherketone has been moulded successfully into
jet engine parts, a good example of how the development of an
engineering thermoplastic is extending the range of use of the more
expensive fibres. These newer and more elaborate thermoplastic
composites provide some exciting prospects for the production of
high temperature engineering components. This book has been
written with these types of material in mind.

The plan of this book is as follows. Chapter 2 discusses the
relevant theory underlining short fibre reinforcement, with a view
to providing a basis upon which the observed mechanical properties
of short fibre composites can be compared. This comparison is
made in Chapter 3 and includes some debate about the rôle of the
fibre-matrix interface in influencing composite properties.

Chapter 4 is concerned with the very practical and important problem of the production of the composite feedstock and the methods of assessing the effectiveness of the compounding operation. In Chapter 5, methods for examining the microstructure of short fibre reinforced thermoplastics are discussed, together with the effects of moulding conditions on the fibre length and orientation in components. The rheological properties of reinforced thermoplastics are discussed in Chapter 6 together with the way in which this information can be used to assist in interpreting the complex pattern of fibre orientation observed in moulded components. Finally, in Chapter 7, the problems of characterizing the mechanical anisotropy in components are discussed, including a critique of the current assessment methods and how alternative testing strategies are emerging.

Reference

Kelly, A. (1973). Strong Solids. Oxford University Press, London.

CHAPTER 2
Theoretical Background

It is not the intention of this chapter to give a comprehensive
coverage of the mechanical properties of fibre reinforced composites.
This topic is already well covered in a number of now familiar
texts - see e.g. Kelly (1973), Piggott (1980). What this particular
chapter attempts to do is to provide some relevant theoretical
background that highlights the rather special problems associated
with short, as distinct from the traditional long fibre composites.
The prediction of the mechanical properties e.g. the stiffness of
a fully aligned long fibre reinforced composite is a difficult
mathematical problem, especially when the load is not applied along
the fibre axis. To perform the same exercise in a short fibre
reinforced composite is even more difficult. The first problem
involved stems from the fact that the fibres cannot be regarded as
infinitely long - normally a convenient mathematical assumption.
In addition, a real moulded component exhibits a very complex
distribution of fibre orientations, which can itself vary from one
point in the moulding to another - see e.g. Fig. 2.1. Any realistic
predictive work has to include the effect of fibre length/length
distribution and a distribution of fibre orientations with respect to
the applied load. If such predictive work is to be of any practical
use, some mathematical rigour must therefore be sacrificed to enable
comparatively rapid calculations to be made. It is the stiffness
of short fibre reinforced thermoplastics that has received most
attention and with reason can be predicted fairly accurately.

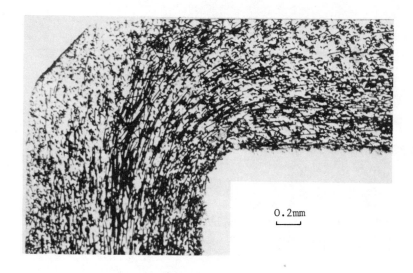

FIG. 2.1.   Optical micrograph of a thin section taken from a carbon
fibre reinforced nylon 66 moulding.

On the other hand, strength and toughness are difficult quantities
to predict even for a long fibre composite, while the corresponding
problem with short fibre reinforced thermoplastics has not yet
received very much attention.   Further complications that can exist
in a reinforced thermoplastic may arise from the presence of a
significant interfacial layer between the fibre and matrix together
with molecular orientation in the matrix phase.

2.1.   ANISOTROPY OF MECHANICAL PROPERTIES

A material is referred to as being anisotropic if the properties
of that material depend on the direction in which they are measured
with respect to some fixed axis.   For example, a fully aligned fibre
reinforced composite will exhibit a much larger Young's modulus when
measured along the fibre direction compared to that obtained when the

the load is applied at 90° to the fibre axis. In most fibre reinforced composites of practical interest, however, the properties do not change if measurements are made in different directions in a plane transverse to the fibre axis (or anisotropy axis). Referring to Fig. 2.2 , this implies that as the angle θ is varied, the properties of the composite will change but that measurements performed at different angles in the 1-2 plane will not reveal any such changes. In this case, the 1-2 plane is referred to as a plane of <u>transverse</u> <u>isotropy</u>.

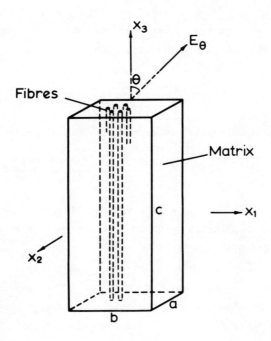

FIG. 2.2. Definition of axes used to describe mechanical anisotropy.

To characterise fully the anisotropy of a composite would seem to
require a very comprehensive set of measurements performed at a
series of angles $0 < \theta < 90^\circ$. For properties such as strength or
toughness this is largely true, but if elastic behaviour only is
being studied then the variation of stiffness with angle θ follows
a definite fundamental relationship:-

$$\frac{1}{E_\theta} = S_{33} \cos^4 \theta + (2S_{13} + S_{44}) \sin^2 \theta \cos^2 \theta + S_{11} \sin^4 \theta$$

$$\dots\dots\dots\dots (2.1)$$

where E_θ is the stiffness measured at angle θ with respect to the
x_3 axis. The S's are compliances and are related to more familiar
elastic properties of the composite, thus:-

$$S_{33} = \frac{1}{E_o} \quad ; \quad S_{11} = \frac{1}{E_{90}} \quad ; \quad S_{44} = \frac{1}{G}$$

where G is the longitudinal shear modulus. S_{13} is given by

the Poisson's ratio $\quad \nu_{13} = -\dfrac{S_{13}}{S_{11}} = \dfrac{\text{Transverse strain along } x_3}{\text{Longitudinal strain along } x_1}$

ν_{13} is likely to be a small quantity, since the transverse strain
along x_3 will be inhibited by the restraining effect of the fibres.
Hence E_θ is effectively defined if values for E_o, E_{90} and G are
known. These quantities may be obtained experimentally or predicted
from a knowledge of the fibre and matrix stiffness and fibre volume
fraction. It is when the latter approach is adopted that the real
complexity of short fibre reinforced composites becomes manifest.
The quantity E_o can be easily predicted with good accuracy for a
long fibre composite but the situation is much less satisfactory
for short fibre composites. E_{90} and G are very difficult to predict
easily and accurately for both types of composite. The next section
will indicate how estimates of all these quantities may be obtained.

The above discussion has been made with reference to a fully aligned
fibre reinforced composite, but the variation of E_θ with θ is also

applicable to a partially oriented system having a single axis of anisotropy, providing the appropriate values of the compliances are used for that particular orientation distribution. Relating the elastic properties of a partially oriented system to those in the fully aligned case requires some assumptions to be made concerning the state of stress and strain throughout the composite. One approach is to assume a state of uniform strain, which gives rise to the Voigt average or to assume a state of uniform stress, which gives rise to the Reuss average. These calculations provide upper and lower bounds for the elastic properties of the partially oriented composite. The reader is referred to the work of Cox (1952), Ward (1962) and Brody and Ward (1971). Simplifications can be made to reduce the computations involved and one popular approach is to ignore the transverse stiffness E_{90} and shear modulus G i.e. to evaluate the effective stiffness of the partially oriented composite by appropriate averaging of E_o over the fibre orientation distribution – Krenchel (1964).

2.2. REINFORCEMENT USING SHORT FIBRES

The stiffness of a composite arises primarily from E_o and so it is necessary to examine the way in which this quantity depends on the properties of the fibre and matrix and the fibre length. When a load is applied along the anisotropy axis of a fully aligned fibre reinforced composite, the load is transferred to the stiff fibres via shear stresses at the interface. The calculation of the variation of shear stress and tensile stress along the fibres was reported by Cox (1952) for the case of an elastic matrix and elastic fibres. This is the now classic "shear-lag" analysis. With reference to Fig. 2.3 consider a fibre of length ℓ and radius r embedded in a matrix. We assume that the matrix as a whole is strained homogeneously by the application of a load applied parallel to the fibre axis. Since the shear stress τ at the fibre-matrix interface will vary along the fibre, so also will the tensile stress in the fibre. Considering a small element of the fibre length δx, the net tensile load δF across this element must be balanced by the shear force

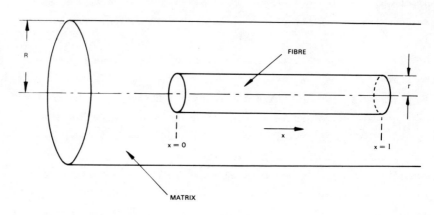

STRESS DIRECTION

R

FIBRE

r

x = 0

x

x = l

MATRIX

FIG. 2.3.   Definition of symbols used in the Cox shear-lag analysis.

at the interface i.e.

$$\delta F = \tau . 2\pi r \delta x \qquad \text{or} \qquad \frac{dF}{dx} = 2\pi r \tau \quad \ldots\ldots (2.2) \text{ when } \delta x \rightarrow 0$$

Since the assumption has been made that the matrix and fibres are
strained elastically, then $\tau \propto \gamma$ where γ is the shear strain developed
at the fibre-matrix interface. Furthermore, γ can reasonably be
expected to be proportional to the difference in elastic displacement
u in the fibre at some point x from one end and the elastic
displacement v of the matrix at the same point, if the fibre were
absent. Hence we may write:-

$$\frac{dF}{dx} = H (u-v) \quad \ldots\ldots\ldots (2.3) \quad \text{where H is a constant for a}$$

particular composite

From Hooke's Law $F = E_f A_f \dfrac{du}{dx}$ $\ldots (2.4)$ where E_f = Young's
modulus of the fibre

A_f = Area of cross-section
of a fibre

Differentiating equation (2.3) with respect to x we have:-

$$\frac{d^2F}{dx^2} = H\left(\frac{du}{dx} - \frac{dv}{dx}\right)$$

but $\frac{du}{dx} = \frac{F}{E_f A_f}$ from equation (2.4), while $\frac{dv}{dx} = \varepsilon$, the strain in the matrix.

hence $\frac{d^2F}{dx^2} = H\left(\frac{F}{E_f A_f} - \varepsilon\right)$

This second order differential equation has a solution of the form:-

$$F = E_f A_f \varepsilon + B \sinh \beta x + C \cosh \beta x$$

where $\beta = \left(\frac{H}{E_f A_f}\right)^{\frac{1}{2}}$ and B and C are constants of

integration, whose values are determined by two boundary conditions.
These are that no loads are transferred across the end faces of a
fibre i.e. $F = 0$ at $x = 0$ and $x = \ell$. We then obtain for the
distribution of tensile stress along the fibre:-

$$\sigma_f = \frac{F}{A_f} = E_f \varepsilon \left\{ 1 - \frac{\cosh \beta(\ell/2 - x)}{\cosh \beta\ell/2} \right\}$$

From this result, the average tensile stress developed in the fibre
can be easily evaluated and is given by:-

$$\bar{\sigma}_f = E_f \varepsilon \left\{ 1 - \frac{\tanh \beta\ell/2}{\beta\ell/2} \right\}$$

The average longitudinal stress in a composite containing a volume
fraction of fibres, v, can then be calculated as a weighted average
of the stresses developed separately in the fibre and matrix i.e.:-

$$\sigma_o = v \bar{\sigma}_f + (1-v)\sigma_m \text{ where } \sigma_m \text{ is the stress developed in the}$$
$$\text{matrix}$$

But since $\sigma_m = E_m \varepsilon$, where E_m = Young's modulus of the matrix, we have:-

$$\sigma_o = v\, E_f\, \varepsilon \left\{ 1 - \frac{\tanh \beta\ell/2}{\beta\ell/2} \right\} + (1-v)E_m\, \varepsilon$$

and hence the effective longitudinal modulus of the composite is given by:-

$$E_o = \frac{\sigma_o}{\varepsilon} = v\, E_f \left\{ 1 - \frac{\tanh \beta\ell/2}{\beta\ell/2} \right\} + (1-v)E_m \quad \ldots\ldots\ldots (2.5)$$

If the mean centre-to-centre separation of the fibres normal to their length is R, then Cox (1952) showed that $H = 2\pi G_m / \log_e\left(\dfrac{R}{r}\right)$

where G_m is the shear modulus of the matrix.

Hence $\beta = \left\{ \dfrac{2\pi G_m}{E_f A_f \, \log_e \left(\dfrac{R}{r}\right)} \right\}^{\frac{1}{2}}$

Now obviously R and r must be related to the volume fraction of fibres in the composite. If we assume that the fibres are arranged hexagonally, then it can easily be shown that the volume fraction of fibres $v = \dfrac{2\pi r^2}{\sqrt{3}\, R^2}$

From these equations, we see that for any given composite, the value of β may be evaluated given values for G_m, E_f, A_f and v. The magnitude of β will essentially determine the "scale" of the dependence of E_o on fibre length ℓ e.g. it will directly affect the length of fibre required to give a value of E_o close to that expected for an infinitely long fibre composite. The predicted dependence of E_o on fibre length ℓ for two fibre reinforced thermoplastics, of current commercial interest, are given in Fig. 2.4. It is clear that if composite stiffnesses are required which approach the maximum, then fibres having lengths ~ 1-2mm are required. This is particularly important in the case of expensive fibres such as carbon, otherwise their advantage compared with cheaper competitors will not be realised.

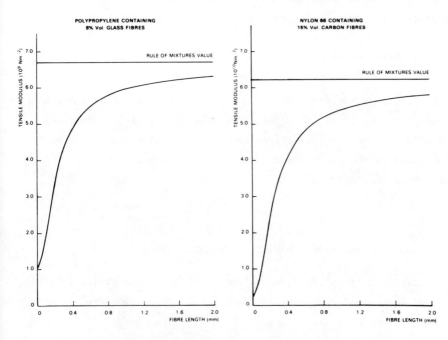

FIG. 2.4. Theoretical dependence of the longitudinal stiffness on
fibre length in aligned short fibre reinforced thermoplastics.

The analysis presented above is for the case where both the fibre
and matrix exhibit elastic behaviour. Most thermoplastics used
as matrix systems, however, will probably be ductile i.e. they will
show a yield point under tensile and shear loadings. If this is
the case, then the shear stresses developed towards the fibre ends
can exceed the shear yield strength, τ_y, of the matrix, even for
small composite strains. The effect is to limit the maximum shear
stress at the fibre-matrix interface to τ_y. Whereas in the Cox
analysis the variation of tensile stress towards the fibre ends is
strictly non-linear, the presence of a constant shear stress τ_y
will lead to a simple linear increase in tensile stress from the
ends of the fibre until a constant tensile stress of $E_f \varepsilon$ is reached.
This analysis was originally developed by Kelly and Tyson (1965) for

metallic matrices having very low yield stresses. The dependence of E_o and ℓ will be essentially similar to that using the Cox approach, the only difference being in the effective value of β. The simpler dependence of fibre tensile stress on x given by the Kelly - Tyson model is of particular benefit when composite strength rather than stiffness is being considered.

The transverse tensile modulus E_{90} and longitudinal shear modulus G are very difficult to calculate from first principles. However, as with continuous fibre composites, both of these quantities are dominated by the matrix phase and may be estimated using the following relationships:-

$$\frac{1}{E_{90}} = \frac{v}{E_f} + \frac{(1-v)}{E_m} \quad ; \quad \frac{1}{G} = \frac{v}{G_F} + \frac{(1-v)}{G_m}$$

where the subscripts have their usual meaning.

In most short fibre reinforced thermoplastics of commercial interest, $E_f \gg E_m$ and $G_f \gg G_m$ in which case $E_{90} \sim E_m/(1-v)$ ie. close to the tensile modulus of the matrix.

2.3 THE STIFFNESS OF PARTIALLY ORIENTED COMPOSITES

The discussion so far has been concerned with the mechanical properties of uniaxially aligned short fibre composites. In moulded components, however, the fibres are rarely perfectly aligned and so unless some provision is made in theoretical analyses to allow for partially oriented systems, predictive work will be of little practical value. There are currently at least three approaches that may be considered:-

(i)   Evaluation of the relative efficiency of reinforcement by fibres in each orientation followed by an averaging procedure over the whole composite - Krenchel (1964), Cox (1952).

(ii) Treating the composite as an assembly of laminates. The stiffness of each layer is predicted or measured experimentally and the stiffness of the "plies" are summed to give the stiffness of the composite - Jerina et al (1973).

(iii) The composite is regarded as an aggregate of sub-units, each sub-unit possessing the elastic properties of a reinforced composite in which the fibres are continuous and fully aligned. Upper and lower bounds for the composite elastic constants are then obtained by averaging the elastic constants for the sub-unit, using orientation functions based on Legendre polynomials - Brody and Ward (1971).

Of these methods, the latter, although perhaps the most comprehensive, would require protracted computations. The first two techniques on the other hand are comparatively straightforward to apply and have proved successful in interpreting the anisotropy in simple moulded components, as discussed later in Chapter 7.

Krenchel's approach was originally developed to account for the stiffness and strength of fibre reinforced cement. Taking the case for the moment of uniaxially aligned continuous fibres, it is possible to assign an orientation efficiency factor η_o in order to account for the anisotropy of stiffness. Thus the composite stiffness would be given by:-

$$E = \eta_o \, v \, E_f + (1-v)E_m$$

where e.g. $\eta_o = 1$ if the load were applied parallel to the fibre axis and $\eta_o = 0$ if it were perpendicular. If the load is applied at some angle θ with respect to the fibre axis then $\eta_o = \cos^4 \theta$. These results are based on the assumption that the fibre and matrix phases suffer the same strain at any given angle θ i.e. that a Voigt average is applicable. If the composite is partially oriented, then η_o is determined by dividing the reinforcement into groups of uniaxially aligned fibres in which case we have:-

$$\eta_o = \sum_n a_n \cos^4 \theta_n \quad \text{where} \quad \sum_n a_n = 1$$

and a_n is the proportion of those fibres making an angle θ_n with respect to the applied load. If the reinforcement cannot be divided into groups of parallel fibres, the summation is made as an integration.

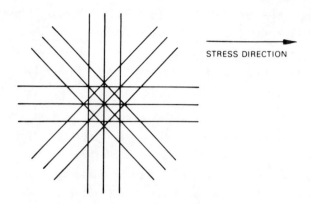

FIG. 2.5. Simple arrangement of fibres used in the calculation of a reinforcement efficiency factor.

As a simple example of the Krenchel approach we could take four equal groups, each consisting of parallel fibres and arranged as shown in Fig. 2.5. In this case we would have:-

$$a_1 = a_2 = a_3 = a_4 = \tfrac{1}{4}$$

$$\theta_1 = 0, \theta_2 = \pi/2, \; \theta_3 = \pi/4, \; \theta_4 = {}^{-\pi}/4$$

Hence $\eta_o = \tfrac{1}{4} (1 + 0 + \tfrac{1}{4} + \tfrac{1}{4}) = \tfrac{3}{8}$

Further examples of calculated efficiency factors for other idealised geometries are shown in Fig. 2.6. Since we are concerned with short fibres, the efficiency factor will be lower than that given above. The total efficiency factor may be written $\eta = \eta_o \eta_\ell$ where $\eta_\ell = 1 - \dfrac{\tanh \beta\ell/2}{\beta\ell/2}$ if the Cox analysis is used, in which case the stiffness of the composite is given by:-

$$E = \eta_o \; \eta_\ell \; v \; E_f + (1-v)E_m$$

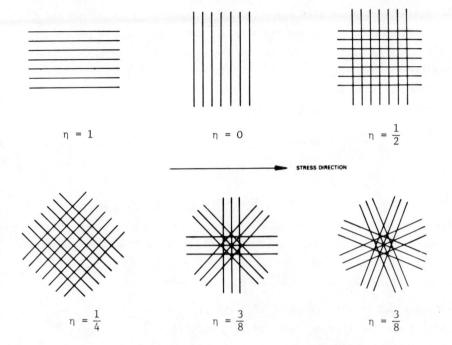

$\eta = 1$ $\eta = 0$ $\eta = \frac{1}{2}$

STRESS DIRECTION

$\eta = \frac{1}{4}$ $\eta = \frac{3}{8}$ $\eta = \frac{3}{8}$

FIG. 2.6. Examples of calculated reinforcement efficiency factors for various idealised arrangements of fibres.

If the fibres are not all of one length then either the distribution of fibre lengths can be represented by a single number e.g. the number average fibre length or η_ℓ must be obtained by summing the effects over the entire population of fibres.

The approach adopted by Halpin and coworkers is based on a theory for predicting the stiffness of a laminate consisting of a number of individual uniaxially aligned plies oriented at known angles to a reference axis. For an individual ply the fibres may be either continuous or discontinuous and the properties of the ply may be determined using any of the methods discussed in this chapter or by use of the Halpin-Tsai equations - see e.g. Ashton et al (1969). Futhermore, Jerina et al (1973) have proposed that short fibre reinforced composites having an essentially planar (two-dimensional) fibre orientation distribution may be modelled mathematically as laminates. Thus the thickness of a "ply" t_n aligned at angle θ_n to the reference axis in the <u>real</u> laminated composite is considered proportional to the fibre fraction a_n having an angle θ_n in the fibre orientation distribution of the short fibre composite. The stiffness of the composite is calculated by summing the stiffness of the "fictitious" plies, using well established methods developed for laminated composites i.e.:-

$$E = \sum_n \frac{t_n}{t} E_n \text{ where } t = \sum_n t_n$$

and E_n is the stiffness of the n'th ply.

There may appear to be very little difference between the Krenchel and Halpin methods as far as the form of the final equations is concerned, but note that in the former an "overall" efficiency factor is being evaluated, whereas the latter sums the stiffnesses of individual fractions of fibres aligned at angle θ_n.

The predictive work of Cox (1952) and Brody and Ward (1971) is similar to the approach used by Krenchel (1964) in that the partially oriented composite is considered as an aggregate of sub-units consisting of fully aligned fibres, but with the symmetry axes of these sub-units arranged according to some orientation distribution function. These theories require a knowledge of most of the elastic constants of the sub-units. Brody and Ward (1971) evaluate these using series/parallel models, assuming that the fibres are infinitely long. The algebra is tedious, although straightforward, and the reader is referred to the original papers for more details.

2.4. STRENGTH OF SHORT-FIBRE COMPOSITES

Prediction of the strength of short fibre reinforced thermoplastics is a complex but industrially crucial problem. Even in the case of unidirectionally aligned fibres with the tensile stress applied along the fibre axis, failure may occur in the fibres, in the matrix phase or at the fibre/matrix interface. Most fibres used in thermoplastics have a much smaller elongation to break than the parent matrix. If therefore the fibres are well bonded to the matrix and are very long the failure stress σ_{uc} of the composite should be given by:-

$$\sigma_{uc} = v\,\sigma_{uf} + (1-v)\sigma_m'$$

where σ_{uf} is the tensile strength of the fibre and σ_m' is the stress carried by the matrix at the fibre failure strain.

In practice, the strength of a continuous fibre reinforced composite rarely approaches the value predicted by this equation. Furthermore, in the case of a composite containing short fibres, the existence of a non-uniform stress along the fibres implies that the average stress $\bar{\sigma}_{uf}$ carried by the fibres at failure will be less than σ_{uf}. The actual relationship connecting $\bar{\sigma}_{uf}$ and σ_{uf} will of course depend on the exact form of the stress distribution at the fibre ends. This has already been assessed for the case of an elastic matrix and elastic fibres using the Cox analysis. When the matrix is ductile and large composite strains are being considered, a linear variation of tensile stress in the fibre with distance from the fibre ends can be used without great loss of generality (Kelly and Tyson 1965). In this case:-

$$\bar{\sigma}_{uf} = \left(1 - \ell_c/2\ell\right)\sigma_{uf} \quad \text{for } \ell > \ell_c \text{ and}$$

$$\ell_c = \frac{\sigma_{uf}\,d}{2\,\tau_u}$$

where d is the fibre diameter and τ_u is the shear strength of the interface or of the matrix, whichever is the weaker. ℓ_c is referred to as the CRITICAL FIBRE LENGTH. The composite strength is then given by:-

$$\sigma_{uc} = \sigma_{uf}\,v\left(1 - \frac{\ell_c}{2\ell}\right) + (1-v)\sigma_m' \quad \cdots\cdots\cdots \text{ (2.6) for } \ell > \ell_c$$

If the fibres are shorter than ℓ_c, the maximum fibre stress is only $2\tau_u\,\ell/d$ and since the mean stress will be half of this, the composite strength will be:-

$$\sigma_{uc} = \frac{\tau_u\,\ell}{d}\,v + (1-v)\sigma_m' \quad \text{for } \ell < \ell_c$$

In most short fibre reinforced thermoplastics, however, there will be a distribution of fibre lengths and then it is necessary to sum the contributions to the composite strength arising from fibres of sub-critical and super-critical length:-

$$\sigma_{uc} = \left\{ \sum_i \frac{\tau_u \ell_i v_i}{d} + \sum_j \sigma_{uf} v_j \left(1 - \frac{\ell_c}{2\ell j} \right) \right\} + (1-v)\sigma_m' \ldots \ldots (2.7)$$

ℓ_c may be evaluated easily, providing values for σ_{uf}, d and τ_u are known for any particular fibre-matrix combination. For carbon fibres in nylon 66, ℓ_c ~100μ, while for well bonded glass fibres in nylon 66, ℓ_c ~230μ. After injection moulding, an appreciable proportion of the fibres present in a short fibre reinforced thermoplastic will have $\ell < \ell_c$ (in some cases the fraction can be as high as 80%) and so it is not surprising that the strengths of a large number of moulded components fall well short of the values expected for continuous reinforcement.

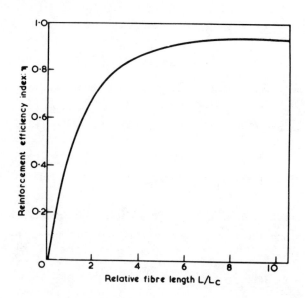

FIG. 2.7. Relationship between relative fibre length and reinforcement efficiency factor η for composite strength. For an ideal composite reinforced with continuous aligned fibres, η = 1. (After Bader and Bowyer, 1973 - reference, Chapter 4)

As with stiffness, the strength of an aligned short fibre
reinforced thermoplastic increases monotonically as the fibre length
increases and approaches an asymptotic value corresponding to a long
fibre composite - see Fig. 2.7. On the basis of the simple analysis
presented above, it would seem that some 95% of the strength of the
continuous fibre material can be developed in a short fibre composite
provided $l/l_c > 10$. This result, however, is over optimistic since
there are other factors that further limit the strength of short
fibre composites. First, it has been shown that the ends of the
fibres act as notches and generate considerable stress concentrations
which can initiate matrix failure - Curtis et al (1978). Confirmation
of this has been obtained using acoustic emission techniques. In
addition to this, the interaction between neighbouring fibres appears
to constrain matrix flow i.e. the matrix becomes embrittled due to
the presence of the fibres. Quite separately, Chen (1971) utilised
finite element methods to confirm theoretically that interacting
fibre stress concentrations reduces the maximum achievable strength
of a short fibre composite to as low as 55% of that expected for
long fibres. Although concerned with epoxy resin matrices, it is
interesting to note that Masoumy et al (1980) have shown that the
fracture process occurring in an aligned short glass fibre composite
involves matrix fracture between bundles of fibres as distinct from
individual fibres. They suggest that it is the bundle and not the
individual fibre aspect ratio that is relevant in predicting the
strength of a short fibre composite. If this result holds for
reinforced thermoplastics, then it would be a possible reason for the
unexpectedly large values of l_c which are frequently found in many
short fibre composites, including those having nominally well bonded
fibres.

As with stiffness, the strength of an aligned short fibre composite
decreases as the angle between the fibre axis and loading direction
increases. Indeed, when the load is applied transversely, the strength
of the composite can sometimes be less than the matrix, due to the
stress raising effect of the fibres. Although the anisotropy of
stiffness in a composite is given accurately by equation 2.1, the
situation with strength is far less clear cut. The prediction of

strength anisotropy requires a failure criterion and it is not obvious, a priori, which is the most appropriate for any particular system. In the context of short fibre composites, there are two criteria that have received widespread attention. The first is due to Stowell and Liu (1961) who invoke a maximum stress criterion and define three failure mechanisms:-

(a) For stresses directed along or at small angles θ to the fibre axis, failure will be controlled by the fibre strength, as discussed above.

(b) At larger angles θ, increasingly large shear stresses will develop both in the matrix and along the fibre-matrix interface, so that the dominant failure mode will be due to shear processes.

(c) At very large θ, approaching 90°, the mode will change again to one of transverse tensile failure either in the matrix or at the interface.

These may be expressed mathematically as follows:-

$$\sigma_{u\theta} = \sigma_{uc} \sec^2\theta \text{ for fibre tensile failure}$$

$$\sigma_{u\theta} = 2\tau_{uc} \text{ cosec } 2\theta \text{ for shear failure parallel to the fibres}$$

$$\sigma_{u\theta} = \sigma_{ut} \text{ cosec}^2\theta \text{ for tensile failure normal to the fibres}$$

and $\sigma_{u\theta}$ is the composite strength at an angle θ to the fibres, σ_{uc} and σ_{ut} are the strengths of the uniaxially aligned composite parallel with and normal to the fibres, and τ_{uc} is the in-plane composite shear strength. A schematic diagram showing the variation of $\sigma_{u\theta}$ with θ, based on the Stowell-Liu equations is shown in Fig. 2.8.

An alternative failure criterion which has also been found to fit strength data for short fibre composites is due to Hill (1948) and Azzi and Tsai (1965). They use a maximum distortional energy theory based on the Von Mises failure criterion, which gives the following variation of $\sigma_{u\theta}$ with θ:-

$$\frac{1}{\sigma_{u\theta}^2} = \frac{\cos^4\theta}{\sigma_{uc}^2} + \frac{\sin^4\theta}{\sigma_{ut}^2} + \left(\frac{1}{\tau_{uc}^2} - \frac{1}{\sigma_{uc}^2}\right) \sin^2\theta \cos^2\theta$$

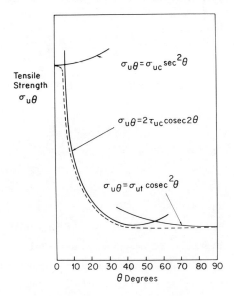

FIG. 2.8. Predicted anisotropy of strength in an aligned fibre
reinforced composite, based on the equations due to Stowell and
Liu, 1961.

The choice between these two approaches can only be made by comparison
with experimental data. This will be discussed in Chapter 3.

Of course, moulded components will not consist of just uniaxially
aligned fibres. Here the approach adopted by Krenchel (1964) and
discussed in section 2.3 can also be applied to obtain an estimate
of the strength of a composite containing misaligned fibres. In
the special case where the fibres are random in a plane, Lees (1968)
integrates the Stowell-Liu equations over all angles θ to give the
following equation for the average strength of the composite:-

$$< \sigma_{uc} > \; = \; \frac{2\,\tau_{uc}}{\pi} \left\{ 2 + \log_e \left(\frac{\sigma_{uc}\,\sigma_{ut}}{\tau_{uc}^{\,2}} \right) \right\}$$

where, as before, τ_{uc} is the composite shear strength and σ_{ut} is the
transverse strength of the uniaxially aligned composite.

Using a similar approach, Chen (1971) concludes that the predictive equation should be:-

$$< \sigma_{uc} > = \frac{2\tau_{uc}}{\pi} \left\{ 2 + \log_e \left(\xi \frac{\sigma_{uc} \, \sigma_m'}{\tau_{uc}^2} \right) \right\}$$

where ξ is a strength efficiency factor, which relates the strength of a unidirectional short fibre composite to the rule of mixtures prediction.

The idea of integrating the Stowell-Liu equations to obtain the average strength of a random planar mat has been criticized by Phillips and Harris (1977). They point out that although in some cases the predicted strength is in accord with that obtained experimentally, nevertheless an assumption has been made that the three mechanisms envisaged by Stowell and Liu (1961) are independent. This may be reasonable in a uniaxially aligned fibre system, but its validity is not obvious, a priori, when the fibres are not uniformly oriented.

More recently, Halpin and Kardos (1978) have shown that the laminate model originally developed for predicting the stiffness of a composite (see section 2.3) may be extended to included composite strength.

2.5. FRACTURE TOUGHNESS

So far we have been preoccupied with a discussion of the stiffness and strength of short fibre reinforced composites. From a practical point of view, however, a composite must also be reasonably tolerant to impact loading i.e. it must be capable of being damaged without undergoing complete failure. For this to happen there must be energy absorbing mechanisms built into the composite. A number of methods might be considered:-

(i)   The use of intrinsically tough matrices, including for example rubber modified polymers.

(ii)   The application of a soft coating to the fibres which will act as an inter-layer after the composite is fabricated. This has been shown to reduce significantly the stress concentrating effect of the fibres, especially under transverse loading.

(iii) Utilization of the energy required to debond the fibres from the matrix and then to pull the fibres completely out of the matrix.

(iv) Use of a weak interface between the fibre and matrix. In this case the triaxial stress system at the tip of an advancing crack causes debonding to occur and a classical crack blunting mechanism takes place, as envisaged by Cook and Gordon (1964).

It is obvious that the conditions leading to (iii) and (iv) are at odds with those required for good stiffness and strength of the composite. In the latter case, the presence of a weak interface would lead to poor load transfer from the matrix to the fibres and so to a low composite strength. In the case of (iii), the energy involved and hence toughness is greatest when the length of the fibres is equal to the critical length ℓ_c. So again it is apparent that maximum strength and maximum toughness cannot be achieved simultaneously, and that composites must be designed for an optimum combination of the desired mechanical properties. The variation of work of fracture with fibre length has been calculated by Cooper (1970) using a model based on (iii) above. Fibres shorter than ℓ_c will be pulled from the matrix, rather than broken, when a crack passes through the composite. The fracture energy will then largely be a combination of the work needed to debond the fibres from the matrix and the work done against friction in pulling the fibres out of the matrix. The fracture energy arising from fibre pull-out is given by the following expressions, due to Cottrell (1964):-

$$U_1 = \frac{v \, \tau \, \ell^2}{12 \, d} \qquad \text{for } \ell < \ell_c$$

where d is the fibre diameter and τ is the interfacial frictional stress.

When $\ell > \ell_c$, only a proportion of the fibres will pull-out and in this case the energy is given by $U_2 = \dfrac{v \, \tau \, \ell_c^3}{12 \, d \, \ell} \qquad \text{for } \ell > \ell_c$.

Hence $U_1 \propto \ell^2$ for $\ell < \ell_c$ and $U_2 \propto 1/\ell$ for $\ell > \ell_c$. The form of the variation of work of fracture with fibre length is shown in Fig. 2.9.

The energy reaches a maximum when $\ell=\ell_c$ and will have a value U_{max} given by:-

$$U_{max} = \frac{v\,\tau\,\ell_c^{\,2}}{12\,d}$$

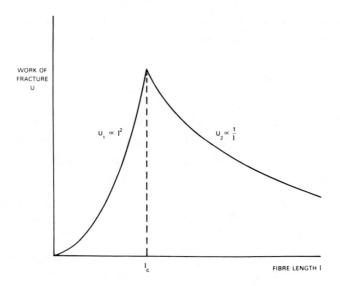

FIG 2.9. Predicted dependence of composite work of fracture on fibre length.

Strictly speaking, to this energy should be added the work of fracture arising from the fibres and matrix, plus the energy to debond the fibres. However, in many cases of practical interest, the energy of pull-out considerably exceeds these other contributions. Note also that since $\ell_c \propto d$, $U_{max} \propto d$. This is of some technological consequence, since it provides one explanation for the experimental observation that the presence of poorly dispersed fibre bundles can significantly increase the impact strength of short fibre reinforced thermoplastics. The fibre bundles are presumably acting as a single large diameter fibre as far as composite fracture toughness is concerned.

2.6. OPTIMISATION OF STIFFNESS, STRENGTH AND TOUGHNESS

From the subject matter discussed in this chapter, it is clear that to achieve high values for the stiffness and strength of a short fibre composite, it is necessary to use long fibres, well bonded to the matrix. For maximum toughness, on the other hand, it is desirable to have a weak interface or use fibres having a length $\ell \leqslant \ell_c$. In a moulded component, of course, there will be a fibre length distribution, and so one can expect a degree of toughness enhancement due to the presence of a proportion of short fibres. However, if the fibres happen to be carbon, for example, it is thoroughly undesirable and uneconomic to use a proportion of them for toughening the composite. In this case, it would be much more advantageous to ensure that the expensive carbon fibres are fully exploited to stiffen and strengthen the composite, while using a much cheaper fibre for toughening purposes. This is the concept of hybrid composites and which has been successfully applied to improve the properties and cost effectiveness of continuous fibre composite components – see e.g. Short and Summerscales (1979,1980).

The use of intimately mixed carbon and glass fibres for improving the overall properties of a short fibre reinforced composite has been advocated by Richter (1977). His results indicate that the change in composite properties corresponds approximately to the change in proportions of the two fibre species. However, both species of fibre were sufficently long (2-4 mm) to ensure effective load transfer from matrix to fibre during composite loading. The observed improvement in work of fracture is probably due to the higher strain to failure of the glass fibres compared to carbon, rather than fibre pull-out. There is a great deal of scope for tailoring composites to specific design requirements by mixing together the various types of fibres that are currently available.

References

Ashton, J.E., Halpin, J.C and Petit, P.H. (1969). Primer on Composite Analysis, Chapter 5. Technomic Publishing Co., Stamford, Conn.

Azzi, V.D. and Tsai, S.W. (1965). Anisotropic strength of composites Exp. Mech., 5, 283-288.

Brody, H. and Ward, I.M. (1971). Modulus of short carbon and glass fibre reinforced composites. Polym.Eng.Sci., 11, 139-151.

Chen, P.E. (1971). Strength properties of discontinuous fiber composites. Polym.Eng.Sci., 11, 51-56.

Cook, J. and Gordon, J.E. (1964). A mechanism for the control of crack propagation in all-brittle systems. Proc.Roy.Soc., A282, 508-520.

Cooper, G.A. (1970). The fracture toughness of composites reinforced with weakened fibres. J.Mat.Sci., 5, 645-654.

Cottrell, A.H. (1964). Strong solids. Proc.Roy.Soc., A282, 2-9.

Cox, H.L. (1952). The elasticity and strength of paper and other fibrous materials. Brit. J. Appl. Phys., 3, 72-79.

Curtis, P.T., Bader, M.G. and Bailey, J.E. (1978). The stiffness and strength of a polyamide thermoplastic reinforced with glass and carbon fibres. J.Mat.Sci., 13, 377-390.

Halpin, J.C. and Kardos, J.L. (1978). Strength of discontinuous reinforced composites: I Fibre reinforced composites. Polym. Eng. Sci., 18, 496-504.

Hill, R. (1948). A theory of the yielding and plastic flow of anisotropic metals. Proc. Roy. Soc., A193, 281-297.

Jerina, K.L., Halpin, J.C. and Nicolais, L. (1973). Strength of molded discontinuous fiber composites. Ing. Chim. Ital., 9, 94-102.

Kelly, A. (1973). Strong Solids. Oxford University Press, London.

Kelly, A and Tyson, W.R. (1965). Tensile properties of fibre-
reinforced metals: Copper/tungsten and copper/molybdenum. J.Mech.
Phys. Solids, 13, 329-350.

Krenchel, H. (1964). Fibre Reinforcement. Akademisk Forlag,
Copenhagen.

Lees, J.K. (1968). A study of the tensile strength of short fiber
reinforced plastics. Polym. Eng. Sci., 8, 195-201.

Masoumy, E., Kacir, L. and Kardos, J.L. (1980). Effect of fiber aspect
ratio and orientation on the stress-strain behaviour of aligned,
short fiber reinforced, ductile epoxy. Report, Materials Research
Laboratory, Washington University, St Louis, Missouri.

Phillips, D.C. and Harris, B. (1977). The strength, toughness and
fatigue properties of polymer composites. Polymer Engineering
Composites, chapter 2, ed. M.O.W. Richardson. Applied Science
Publishers, Barking, Essex.

Piggott, M.R, (1980). Load Bearing Fibre Composites. Pergamon Press,
Oxford.

Richter, H. (1977). Hybrid composite materials with oriented short
fibres. Kunststoffe, 67, 739-743.

Short, D. and Summerscales, J. (Oct 1979). Hybrids - a review:
I Techniques, design and construction. Composites, 215-221.

Short, D. and Summerscales, J. (Jan 1980). Hybrids - a review:
II Physical Properties. Composites, 33-38.

Stowell, E.Z. and Liu , T.S. (1961). On the mechanical behaviour
of fibre-reinforced crystalline materials. J.Mech.Phys. Solids,
9, 242-260.

Ward, I.M. (1962). Optical and mechanical anisotropy in crystalline
polymers. Proc. Phys. Soc., 80, 1176-1188.

CHAPTER 3
Fundamental Studies of Mechanical Anisotropy

The previous chapter was concerned with establishing the mechanics
of fibre reinforced materials, such as will affect our understanding
of the stiffness and strength of short fibre reinforced thermoplastics. The purpose of this chapter is to discuss the experimental
work that has been carried out to confirm or consolidate the basic
principles of short fibre reinforcement. For this purpose, it is
most convenient to study composites having a fairly simple arrangement
of fibres e.g. fully aligned or random in a plane, for which
theoretical predictions have been advanced. It will be clear from
the ensuing discussion that it is very difficult in practice to make
such model composites. Nevertheless, a study of the literature
reveals that a distinction can be made between those studies that
seek to confirm theoretical predictions of the mechanics of short
fibre reinforced thermoplastics and those that seek to predict the
properties of real moulded components, having a very complex fibre
orientation distribution. The latter problem has very special
significance in the prediction of the creep properties of short fibre
reinforced composites. This subject will be discussed separately
in Chapter 7. For the moment, however, we will only be concerned
with the experimental observations of the stiffness and strength of
composites, ideally having a simple arrangement of fibres.

Short fibre reinforced thermoplastics present special problems where
control and prediction of properties are concerned. Aside from the
direct effect of fibre length on stiffness and strength as discussed

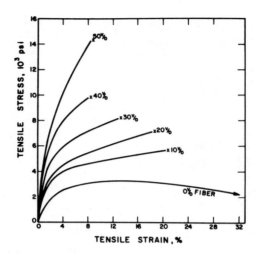

FIG. 3.1. Tensile stress-strain curves for uniaxially aligned
composites of polyester fibre in a polyethylene matrix. Stress
applied parallel to the fibre axis.

(After Blumentritt et al, 1974).

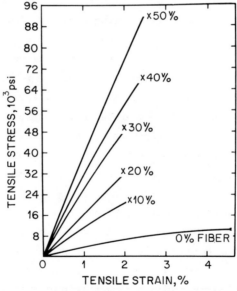

FIG. 3.2 Tensile stress-strain curves for uniaxially aligned
composites of Kevlar 49 fibre in a polymethylmethacrylate matrix.
Stress applied parallel to the fibre axis.

(After Blumentritt et al, 1974)

in the previous chapter, experimental studies have shown that the presence of fibre ends within the body of the composite can cause crack initiation and hence potential composite failure. In addition, the fibres themselves can modify the microstructure of the surrounding matrix by providing nucleation sites along their surface for spherulite growth. In some cases this can produce very significant molecular orientation in the matrix parallel to the fibre axis. The effect of this on composite properties may be considerable, especially when large volume fractions of fibres are being used. Neither of these effects, per se, feature in the study of conventional long fibre composites, most of which are based on thermosets.

Little if any work has been reported, concerned with confirming the predictions of stiffness versus fibre length in well characterised aligned composites. Most of the reported studies concentrate on general mechanical data corresponding to different fibre-matrix combinations or on a characterisation of the anisotropy of stiffness or strength in highly aligned composites.

3.1. TENSILE STRESS - STRAIN BEHAVIOUR

A number of workers have reported the form of the stress-strain curves for various fibre-matrix combinations and volume fractions of fibres - see e.g. Lavengood (1972), Blumentritt et al (1974,1975a), Curtis et al (1978), Weiss (1980). Figs 3.1 and 3.2, taken from the work of Blumentritt et al (1974) show typical stress-strain data for two extreme types of uniaxially aligned composite; one a ductile fibre in a ductile matrix (Fig. 3.1) and the other, a relatively brittle fibre in a brittle matrix (Fig 3.2). Both stiffness and strength are increased by addition of fibres. Ductile matrix composites reinforced with low modulus fibres exhibit a significant reduction in the slope of the stress-strain curve at 2-4% strain. Brittle matrix composites exhibit nearly elastic behaviour to fracture. No major changes in this pattern of behaviour occur for these types of composites if a random planar mat of fibres is used - Blumentritt (1975a). Similar observations on quite highly aligned composites have been made for the commercially important glass and carbon fibre reinforced thermoplastics e.g. polypropylene and nylon - Curtis et al (1978), Weiss (1980). The case of carbon fibres in polypropylene is shown in Fig. 3.3.

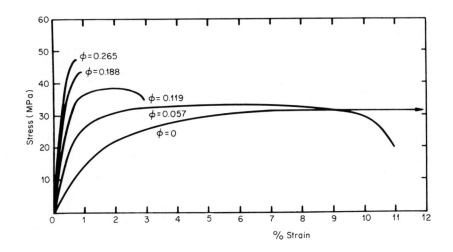

FIG. 3.3. Typical stress-strain curves for carbon fibre reinforced polypropylene at various fibre volume fractions, ϕ. Data obtained from injection moulded tensile bars.
(After Weiss, 1980)

One very important consequence of utilising large volume fractions of fibres is that although stiffness is increased significantly, there is not necessarily a pro rata change in strength. Furthermore, the work to fracture decreases rapidly as the concentration of fibres is increased i.e. the composite will only tolerate small impact energies and will be unable to dissipate any of this energy in plastic flow processes. This is not a problem particular to short fibre thermoplastics but is also present with conventional long fibre composites. This has been one of the motivations behind the development of "hybrid" composites, which attempt to achieve both high stiffness and high work of fracture.

3.1.1 Stiffness

The stiffness of short fibre reinforced thermoplastics depends on
the fibre length (and/or distribution), volume fraction of fibres,
the stress transfer efficiency of the interface and of course the
fibre orientation. The most direct verification of Equation (2.5)
requires a series of composites having fully aligned fibres of
well - defined length. This is difficult to achieve in practice
since methods that give a good fibre dispersion throughout the
matrix usually cause fibre breakage. The only direct study of the
dependence of fibre length on modulus (and strength) was carried
out by Anderson and Lavengood (1968), but this utilized epoxy resins
as matrices. In this case, the pre-chopped fibres were oriented
in a V-shaped channel, impregnated with resin followed by complete
moulding and curing. A typical graph of fibre efficiency (actual
composite modulus ÷ law of mixtures prediction) versus fibre aspect
ratio is shown in Fig. 3.4. The form of this graph is close to
the theoretical prediction (Fig. 2.4) and serves to confirm that a
fibre aspect ratio of at least 100-200 is required in order to
achieve a longitudinal modulus close to that corresponding to an
infinitely long fibre composite. A corresponding study for
thermoplastics has not been reported, but a similar approach would
seem reasonable if in-situ polymerisation of the appropriate monomer
were possible as in the case of nylon 6 - Bessell and Shortall (1975).

Lees (1968a) has studied the anisotropy of the stiffness in highly
aligned composites of short glass fibres in polyethylene, polymethylmethacrylate, a fluorocarbon resin and styrene acrylonitrile. These
samples were prepared by screw extrusion of the fibre/matrix blend
through a circular cross-section die, in order to produce a high
degree of fibre orientation along the extrusion direction. Lengths
of extrudate were cut, stacked side by side and compression moulded.
This produced quite highly aligned plaques from which samples
could be cut with their axes at various angles to the overall
fibre direction. The angular dependence of the stiffness for
aligned fibres in polyethylene and styrene acrylonitrile is shown

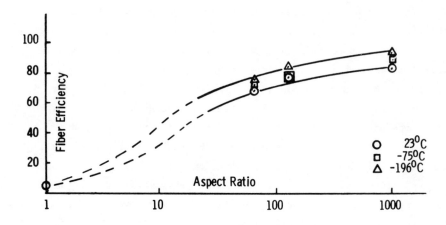

FIG. 3.4. Fibre reinforcement efficiency factor (%) for stiffness
versus fibre aspect ratio, for aligned glass fibres in epoxy
resin.

(After Anderson and Lavengood, 1968)

Figs 3.5 and 3.6. Also plotted is the theoretical curve based on
Equation (2.1). This relationship was fitted to the experimental
data by using values of the stiffness at $\theta = 0^{\circ}$ and 90° together
with a calculated value of S_{44} using the theory due to Whitney and
Riley (1966). The agreement between the theoretical and
experimental angular dependences is excellent. The form of this
relationship is similar to that observed by Lavengood (1972) for
glass fibres in epoxy resins and by McNally (1977) for glass fibres
in polybutylene terephthalate – see Fig 3.7. This figure, in
particular, shows very clearly that the stiffness is a very
sensitive function of angle for small degrees of off-axis loading.
This is a well known fact for uniaxially aligned long fibre composites
and the sensitivity to angle increases as the ratio of the fibre
modulus to the matrix modulus increases, being especially pronounced
for carbon fibre composites. The consequence of this is that unless
the fibres are perfectly aligned, the observed longitudinal stiffness

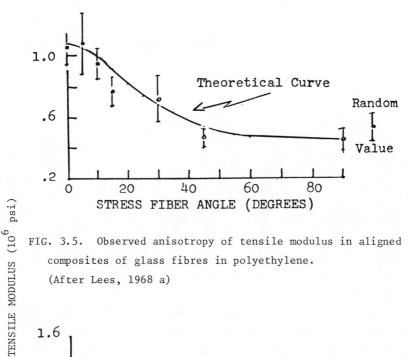

FIG. 3.5.   Observed anisotropy of tensile modulus in aligned
composites of glass fibres in polyethylene.
(After Lees, 1968 a)

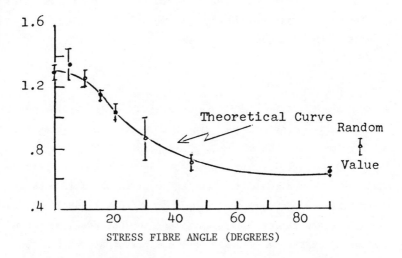

FIG. 3.6.   Observed anisotropy of tensile modulus in aligned
composites of glass fibres in styrene-acrylonitrile.
(After Lees, 1968 a)

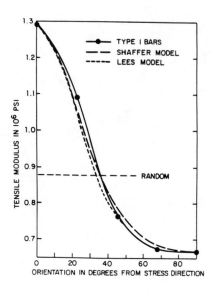

FIG. 3.7. Anisotropy of tensile modulus in aligned samples of glass
fibre reinforced polybutyleneterephthalate.
(After McNally, 1977)

will be substantially less than predicted and the difference in
composite stiffness between using e.g. carbon or glass as the
reinforcement will be minimal. This point has been emphasized by
Brody and Ward (1971) who applied the aggregate model, discussed in
Chapter 2, to published data on the longitudinal stiffness of
short fibre composites, including that of Lees (1968a). In all
cases, the experimental data was close to that predicted using the
continuity of stress (Reuss bound) criterion rather than the
expected continuity of strain (Voigt bound). Preconceived ideas
of longitudinal stiffening would naturally favour the latter bound,
but for the data analysed by Brody and Ward (1971) this does not
appear to be the case, as shown in Fig. 3.8. These conclusions
should be treated with caution, however, since in much of the
published data used by Brody and Ward (1971) no values for fibre

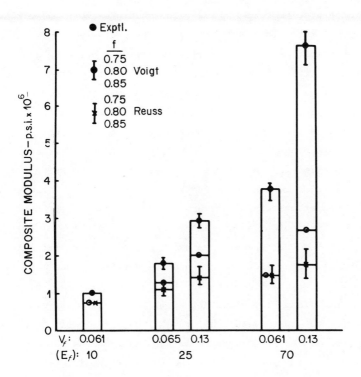

FIG. 3.8. Comparison of the tensile moduli of some fibre reinforced
polymethylmethacrylate composites with predictions based on an
aggregate model.
(After Brody and Ward, 1971)

length are provided and so it is by no means obvious, a priori, that
a Voigt bound would be expected to hold. Furthermore, the
aggregate model will be biassed in favour of the Reuss bound, since
the aggregate "units" are assumed to possess the properties of
an infinitely long fibre composite. In terms of the commercial
exploitation of high modulus fibres in thermoplastics, these
observations may appear distressing since they would suggest that
there is no advantage in using any high performance fibre other than
the cheapest e.g. glass. In practice, of course the situation is
not as bleak as it may appear, since real components will be
subjected to multiaxial loading, where a perfectly aligned short
fibre composite would be inappropriate.

An associated problem concerns the relationship between the stiffness
of an aligned short fibre composite and that possessing a random
fibre orientation. In fact, a composite having a perfect 3D
arrangement of fibres is virtually impossible to fabricate and so
it is probably more realistic to compare the properties of a
uniaxial composite with those of a random planar mat. Such a
composite can be fabricated, (although not easily, if true
randomness in a plane is to be achieved) and has the additional
merit of approximating the actual fibre distribution observed in
parts of injection moulded components – see Chapter 5. A number of
different theoretical approaches have been used to predict the
in-plane stiffness of a random planar mat, knowing the elastic
properties of the corresponding uniaxial system. An experimental
examination of the predictions of various theories can be made by
reference to the results of Lees (1968a), Ogorkiewicz (1971),
Lavengood (1972) and Blumentritt et al (1974, 1975a). Lavengood (1972)
in his work with short glass fibre reinforced epoxy resin finds that
the stiffness of the random system is about two-thirds the
longitudinal stiffness of the unidirectional material. The
corresponding results of Lees (1968a) are shown in Figs. 3.5 and
3.6, showing that the random stiffness is approximately equal to
0.5–0.6 of the longitudinal stiffness. McNally (1977) has pointed
out that one difficulty that can arise with any study such as this is
that quite different methods are often employed to produce the two
types of composite specimen. This problem is partially overcome in
the work of Blumentritt et al (1974, 1975a) who used a compression
moulding method for the production of both the uniaxially oriented
and random planar mat samples. They found that the calculated
values of the in-plane modulus of a random planar mat using equations
proposed by Cox (1952), Krenchel (1964) and Tsai and Pagano (1968)
were all significantly higher than the measured modulus values.
They conclude that perhaps because of defects present in the samples,
composite modulus values cannot be accurately predicted using the
properties of the fibre and matrix alone. However, if the moduli
of the random planar mats used by Blumentritt et al are compared
directly with the experimental values of the longitudinal moduli for

the corresponding uniaxial composites, the results are similar to those of Lees (1968a) and Lavengood (1972). A summary of some of the results obtained by these workers is given in Table 1.

FIBRE	MATRIX	FIBRE VOL. FRACTION %	E_0 (exp) GNm^{-2}	E_0 (calc) GNm^{-2}	E_{RM} (exp) GNm^{-2}	$\dfrac{E_{RM} \text{ (calc)}}{E_0 \text{ (calc)}}$ COX	$\dfrac{E_{RM} \text{ (calc)}}{E_0 \text{ (calc)}}$ KRENCHEL	$\dfrac{E_{RM} \text{ (exp)}}{E_0 \text{ (exp)}}$
GLASS	EPOXY	50	29.00	-	19.3	0.33	0.375	0.67
GLASS	POLYETHYLENE	10	4.47	8.20	2.29	"	"	0.51
"	"	20	9.45	15.33	3.41	"	"	0.36
"	"	30	10.83	22.47	5.02	"	"	0.46
CARBON	"	10	11.03	24.40	4.62	"	"	0.42
"	"	20	19.30	47.74	6.66	"	"	0.35
"	"	40	23.17	94.41	9.31	"	"	0.40
KEVLAR 49	NYLON 12	10	4.97	14.22	3.08	"	"	0.62
"	"	20	8.55	27.20	4.17	"	"	0.49
"	"	30	11.93	40.18	5.37	"	"	0.45

TABLE 1. Experimental and calculated values of the longitudinal and planar mat tensile moduli for a number of fibre-matrix systems. The symbols have the following meaning:-

E_0(exp) =Experimental value of the longitudinal tensile modulus for samples containing aligned fibres.

E_0(calc) =Longitudinal tensile modulus as predicted by the Rule of Mixtures.

E_{RM}(exp) =Experimental value of the in-plane tensile modulus for random planar mat samples.

E_{RM}(calc) =Calculated value of the in-plane tensile modulus for random planar mat samples.

One may conclude from these studies that the ratios of the moduli of the random planar mats to those of the corresponding uniaxial system exceed the predicted values but that the absolute magnitudes of the predicted moduli exceed the experimental values. The latter statement is also applicable to the uniaxial composites studied by Blumentritt et al, encompassing a wide range of matrices, fibres and volume fractions. In view of this it is difficult to decide whether the discrepancy is due to imperfect wetting of the fibres by the matrix materials or to a fibre length effect. In general, therefore, the comparison of theoretical predictions for both

uniaxial and random planar mats with the corresponding experimental
data remains confused.

3.1.2. Strength

As already mentioned, the strength of a short fibre reinforced
thermoplastic is dependent on fibre length, volume fraction of
fibres, the interfacial shear strength and fibre orientation. With
the exception of anisotropic effects, studies of the tensile strength
of short fibre composites are concerned first with its dependence
on fibre volume fraction and then with its quantitative prediction.
A linear dependence of strength on volume fraction of fibres is
expected from Equation (2.6). This has been observed by Lavengood
(1972) and Blumentritt et al (1975a) for volume fractions covering
the range 0-50%. More recently, Ramsteiner and Theysohn (1979)
in a study of unidirectionally oriented glass fibres in nylon 66,
polypropylene and polymethylmethacrylate, observed linearity at
intermediate volume fractions (5-15% approximately) but deviations
from this at low and high concentrations. The discrepancy can be
due to any number of causes but is indicative of a change in the
mechanism of composite failure. In their case, they attribute the
discrepancy at low fibre concentration to matrix embrittlement,
promoted by the stress concentration at the fibre ends. At high
concentrations, mutual interaction between the fibres can result
in loss of fibre strength and excessive fibre breakage. A very
similar dependence of composite strength on fibre concentration has
been observed by Curtis et al (1978) for glass and carbon fibres in
nylon 66. Using acoustic emission techniques, they conclude that
matrix cracking at the ends of fibres develops progressively as the
strain in the composite is increased and is the precursor of composite
failure. From the reported studies, it is probably much more
realistic to anticipate a rather complex dependence of strength on
fibre concentration rather than that given by Equation (2.6), which
refers to one very specific mode of composite failure.

Although the detailed mechanisms of failure in a short fibre
reinforced composite are complex, it has become customary to
interpret the observed longitudinal strength using one of the models

describing composite failure e.g. that due to Kelly and Tyson (1965) or Piggott (1966). From Equation (2.7), the strength σ_{uc} is related to the fibre length distribution and interfacial shear strength τ_u. Hence if these latter two quantities are known, σ_{uc} may be predicted and compared with the experimentally determined value. Alternatively, the value of τ_u can be adjusted to give a fit between the theoretical and experimental strengths. This second approach is reminiscent of the analysis used in Chapter 4 to evaluate compounding methods and their effect on the stress transfer efficiency. Lees (1968b) has reported tensile strength measurements made on samples consisting of aligned glass fibres in polyethylene and polymethylmethacrylate Assuming perfect bonding at the fibre-matrix interface, he found that the calculated strengths considerably exceeded the experimental values even when an improved composite model due to Piggott (1966) was used in the calculations. Much of this discrepancy could be accounted for by internal stresses developed at the interface during the moulding and subsequent cooling of the specimens. An extract of some of Lees data is given in Table 2. Ramsteiner and Theysohn (1979) use the alternative approach of fitting the Kelly-Tyson equation to the linear portion of their strength versus fibre volume fraction plots, from which they drive a value for τ_u. They compare the values obtained for various fibre-matrix combinations and temperatures with the corresponding shear yield strengths of the matrix. This latter quantity was derived from the observed temperature dependence of the tensile yield strength using the Von Mises yield criterion. Some of the data obtained by these workers are shown in Table 3 and Fig 3.9. As the test temperature is varied, the changes in τ_u and the shear yield strength of the matrix manifest themselves as changes in the nature of the fracture surface. First, the matrix embrittles as the temperature is lowered, since the shear yield strength will increased. Secondly, if τ_u is relatively temperature independent, then at a low enough temperature, the matrix shear yield strength will exceed τ_u and no matrix material will adhere to the fibres when fibre pull-out occurs during composite fracture. Fig. 3.10 illustrate these changes for specimen of glass fibre reinforced nylon 6.

FIBRE	MATRIX	FIBRE VOL. FRACTION %	EXPERIMENTAL STRENGTH MNm^{-2}	THEORETICAL STRENGTH MNm^{-2}						
				CALC. USING $\tau_u = \sigma_{um}/2$			CALC. INCLUDING RESIDUAL STRESS			
				τ_u	KELLY	PIGGOTT	τ_u'	KELLY	PIGGOTT
GLASS	POLYETHYLENE	3.9	49.4	14.0	77.9	86.2	6.0	57.9	60.0
		8.3	74.5	14.0	130.0	145.0	6.0	84.8	87.6
		13.4	87.6	14.0	174.0	192.0	6.0	97.9	100.7
		19.4	110.0	14.0	199.0	211.0	6.0	106.0	108.0
		26.5	105.0	14.0	210.0	218.0	6.0	103.0	104.0
	P.M.M.A.	1.9	64.4	30.0	67.2	67.5	14.5	59.5	59.7
		4.9	71.0	30.0	97.2	98.6	14.5	75.2	75.2
		8.1	78.0	30.0	112.0	114.0	14.5	81.4	81.4
		18.2	95.0	30.0	230.0	235.0	14.5	134.0	134.0

TABLE 2. Experimental and calculated values of the longitudinal strength of glass fibre reinforced thermoplastics containing aligned fibres.

τ_u = interfacial shear strength.

σ_{um} = tensile strength of the matrix.

(Data extracted from the work of Lees, 1968 b).

MATRIX	GLASS FIBRE SURFACE TREATMENT	PROPERTY	TEST TEMPERATURE °C			
			-30	23	50	70
NYLON 6	NONE	τ_{um} (MNm^{-2})		45	30	25
		τ_u (MNm^{-2})	49	33	23	16
		ℓ_c (mm)	0.40	0.60	0.90	1.35
	SAME COMMERICAL COUPLING AGENT BUT TWO DIFFERENT SIZES	τ_u	57	44	34	26
		ℓ_c	0.37	0.48	0.60	0.80
		τ_u	68	47	35	28
		ℓ_c	0.20	0.30	0.43	0.55
	SPECIAL SIZE AND COUPLING AGENT	τ_u	83	54	42	36
		ℓ_c	0.18	0.28	0.36	0.42
POLYPROPYLENE	COMMERCIAL COUPLING AGENT AND SIZE	τ_{um}	40	19	14	9
		τ_u	15	7	5	5
		ℓ_c	1.40	3.10	4.00	4.40
POLYPROPYLENE - CHEMICALLY MODIFIED		τ_{um}	42	23	16	11
		τ_u	40	23	18	13
		ℓ	0.50	0.90	1.20	1.60

TABLE 3. Interfacial shear strength (τ_u), matrix shear strength (τ_{um}) and critical fibre length (ℓ_c) for glass fibre reinforced thermoplastics at various temperatures.
(Data extracted from the work of Ramsteiner and Theysohn, 1979)

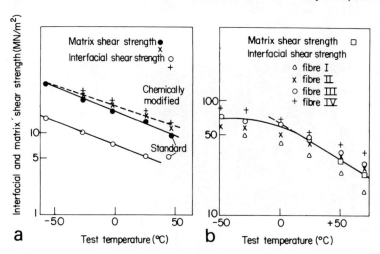

FIG. 3.9. Variation of interfacial and matrix shear strengths with temperature for glass fibre reinforced polypropylene (a) and nylon 6 (b).
(After Ramsteiner and Theysohn, 1979)

46

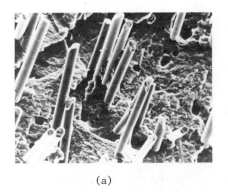

(a)

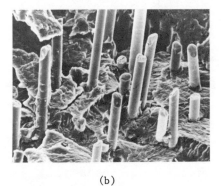

(b)

50µm

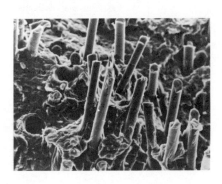

(c)

FIG. 3.10. Scanning electron micrographs of fracture surfaces of
glass fibre reinforced nylon 6 at various test temperatures;
(a) liquid N_2 (b) $-30^{\circ}C$ (c) $+50^{\circ}C$
(After Ramsteiner and Theysohn, 1979)

In the practical utilisation of short fibre composites, it is the
failure processes initiated by off-axis loading that will be much
more important in determining design stresses etc. An assessment
of the anisotropy of the ultimate tensile strength in highly aligned
short fibre composites has been undertaken by a number of workers.
Lavengood (1972) and Masoumy et al (1980) have studied systems of
aligned glass fibres in epoxy resins, while Lees (1968b), McNally
(1977) and Ramsteiner and Theysohn (1979) have studied glass fibre –
thermoplastic systems. Fig. 3.11 shows some results of Lees work
and from this it can be seen that the tensile strength decreases
rapidly as the angle between the mean fibre direction and load exceeds
about 10-20° - a very similar trend in fact to that for stiffness.
The interpretation of the anisotropy data requires a knowledge of
an appropriate failure criterion. In most of the reported work,
the starting point for this is the equations of Stowell and Liu
(see chapter 2) which define a maximum stress theory of failure for
the strength of a continuous aligned composite as a function of
orientation. The predictions make on the basis of this, in the case
of Lees work, are also shown in Fig 3.11. Considering the simplicity
of the approach, the agreement between theory and experiment is
reasonable in general,although this does not appear to be the case
in the results of Ramsteiner and Theysohn (1979). Their anisotropy
data for aligned glass fibres in polymethylmethacrylate fits a
modified Von Mises criterion rather better.

3.1.3. Impact behaviour and work of fracture

The fracture toughness of a composite material is a very important
engineering property. There have been many instances where composite
design has been optimised for stiffness and strength alone, which
can then produce a material that is unable to tolerate impact loading.
The area under the stress-strain curve up to the point of failure,
is a measure of the work of fracture. This is formally related to
the fracture mechanics parameters G_{1C} and K_{1C}, being the critical
strain energy release rate and fracture toughness respectively.
As with many composites, the conditions that lead to high stiffness
and strength also result in low elongation to failure, so that the

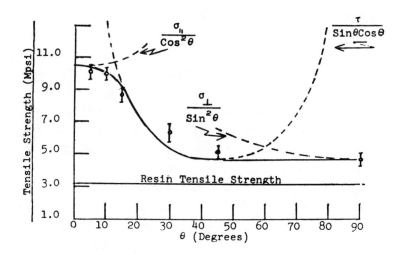

FIG. 3.11.   Anisotropy of tensile strength in aligned composites of
   glass fibres in polyethylene.   Dotted curves are the predictions
   based on the Stowell-Liu analysis.
   (After Lees, 1968 b)

work of fracture can be very low compared to that of the parent
matrix.   The work of fracture is dependent on the presence of energy
dissipative processes and one mechanism in a composite material that
can make a major contribution is the energy required to pull-out
fibres during composite fracture.   As described earlier in the book,
the toughness is maximised when the fibres are at their critical
length, but this condition is not compatible with that required for
high stiffness and strength of the composite.

There has been very little detailed work carried out aimed at
examining the critical parameters affecting the impact strength of
short fibre reinforced thermoplastics.   However, the recent work of
Ramsteiner and Theysohn (1979) illustrates a number of important
points.   If inherently brittle fibres are added to an otherwise
ductile matrix, the impact strength of the composite decreases
rapidly as the fibre concentration increases - see Fig 3.12.
This effect occurs unless the ductility of the matrix is suppressed,

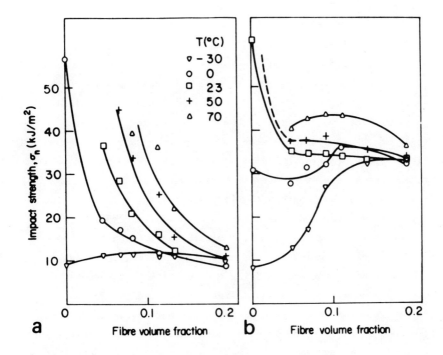

FIG. 3.12. Variation of impact strength with fibre volume fraction at various test temperatures for (a) standard polypropylene (b) polypropylene, which has been chemically modified to promote good fibre-matrix bonding.

(After Ramsteiner and Theysohn, 1979)

as will take place at low temperatures. In this case it appears that the addition of fibres can lead to an increase in impact strength. The reasons for these patterns of behaviour are not difficult to understand. The addition of brittle fibres to a ductile matrix will reduce the elongation to break of the composite and yet at the same time the contribution to the work of fracture resulting from fibre pull-out will be too small to compensate. At high volume fractions of fibres, when the ductility of the matrix is reduced, due to the constraining effect of neighbouring fibres, those fibres having a length $\ell < \ell_c$ will contribute substantially to the work of fracture. There is then a tendency for the impact strength of the composite as a whole to be comparatively independent of fibre volume fraction. A similar mechanism will occur at moderate fibre concentrations at low temperatures, where presumably the increase in impact strength with fibre concentration is again due mainly to those fibres having a length $\ell < \ell_c$. The detailed shape of the plots shown in Fig 3.12 will clearly be dependent on the fibre length distribution in the composite. Indeed, there is considerable scope for manipulating the relative magnitudes of strength and toughness in a short fibre reinforced thermoplastic by appropriate choice of fibre length distribution, which in turn will depend on the particular compounding/moulding route that is used to fabricate the component. Alternative procedures for optimisation of strength and toughness using two or more different species of fibre in any one composite have been discussed in section 2.6. Fig. 3.12 also shows that modifications to the interfacial bond strength (either by chemical modification of the fibre surface or the matrix or both) also have significant effects on the impact strength. Without a detailed knowledge of the particular fibre length distributions used in Ramsteiner and Theysohn's work, it is impossible to say whether the increase in impact strength due to chemical modification of the matrix is due to the improved load transfer to the fibres or the greater interfacial frictional forces involved during fibre pull-out.

The utilisation of fibre pull-out energy is only one of a number of possible methods whereby the impact strength of short fibre reinforced thermoplastics may be enhanced. An alternative method that has been considered, is to use rubber modified polymers as matrix materials, which have been specifically developed to give high impact performance. Short fibre composites will exhibit a high strength parallel to the fibre axis, but are comparatively weak under transverse loading. The use of toughened polymers could in principle inhibit crack propagation parallel to the fibres. However, although higher composite toughnesses are obtained compared to those composites using conventional matrices, the full toughness enhancement is by no means exploited. The reason for this may be associated with the presence of a "damage zone" at the tip of a crack in any visco-elastic material. The "radius" of this zone can be $\sim 300\mu$ in rubber toughened polymers and is crucial to the whole mechanism of toughening. However, the thickness of matrix between neighbouring fibres is ~ 10-20μ in a typical short fibre composite and so the damage zone at the crack tip is severely distorted, causing a concomitant decrease in fracture resistance. This is a very similar problem to that discussed by Bascom et al (1977) concerning the fracture reliability of adhesively bonded joints, when the bond line thickness becomes comparable to the damage zone dimensions.

We have seen that the nature of the bond between fibre and matrix has a significant effect on the impact strength. This observation can be further extended through the use of a special interlayer between the fibre and matrix whose properties and thickness are deliberately chosen to maximise composite toughness. This we now discuss in the following section.

3.2. FIBRE - MATRIX INTERFACE EFFECTS

In short fibre reinforced epoxy resin systems, the use of a thin layer of a soft deformable material around the fibres has been shown to reduce the local stresses around the fibres - Arridge (1975) and Marom and Arridge (1976). This is particularly important since it is the local stress concentrations that can initiate composite failure, especially under transverse loading conditions. For continuous fibre composites there should be little associated change in the longitudinal composite strength when flexible interlayers are used compared to uncoated (but chemically surface treated) fibres. For short fibre composites however, the presence of an interlayer will obviously affect the critical fibre length for pull-out, since the shear strength of the interface is significantly reduced. Broutman and Agarwal (1974) have shown that when the load is applied along the fibre axis in a short fibre composite, the modulus of the interlayer has a profound effect on the stress transfer under these conditions, and that a modulus may be found which optimises the fracture toughness of the composite. The work of Peiffer (1979), using latex coated glass fibres, has confirmed that the impact strength of a composite is a function of both the interlayer thickness and interlayer glass transition temperature Tg. Fig 3.13, taken from this work, shows the dramatic improvement in impact strength when the interlayer is rubbery and has a thickness of about 0.2μ. The impact strength under these conditions is about a factor of six greater than that obtained without the use of an interlayer, so there is considerable technological advantage in pursuing this approach.

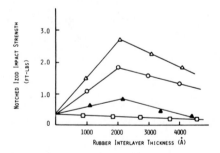

FIG. 3.13. Impact strength of glass fibre reinforced composites possessing interlayers of various thicknesses and having the following glass transition temperatures (^{o}C): -56 (Δ), -14(o), 10 (▲), and 80 (□).
(After Peiffer, 1979)

Such is the situation for reinforced thermosets. In reinforced thermoplastics, however, the possible use of interlayers specially applied to the fibres for the specific purpose of optimising impact properties of the composite has not really been considered. There is of course less incentive, since many of the thermoplastic matrices are at least partially ductile, so that dissipation of crack energy by plastic deformation is much more significant compared to thermoset systems. Nevertheless, fibre manufacturers frequently apply a size to the fibres immediately after fibre production. The function of this from the point of view of the manufacturer is to aid handling of the fibre during composite fabrication and perhaps to improve the wetting of the fibres when incorporated in the thermoplastic. Although this size may have a comparable thickness to normal inter-layers it is debatable whether it functions as a distinct layer after the composite has been fabricated. The reason for this is that the size is often a thermoplastic itself, so that it may melt and form a dilute blend with the matrix phase.

Apart from the possible rôle of the size in influencing composite properties, there is another very major effect that occurs when fibres are incorporated in a thermoplastic, namely the direct influence of the fibre surface on the morphology of the surrounding matrix. Studies of the spherulitic growth of nylon 6 around various types of fibre have been reported by Bessell and Shortall (1975). Observations were made of the crystallisation processes around both single fibres and in composites containing 15% volume fraction of uniaxially aligned fibres. In both cases the matrix was prepared by the in situ anionic polymerisation of caprolactam. It was found that columnar spherulitic growth (frequently referred to as transcrystallinity) occurred around the fibres to a distance of about one or two fibre diameters. Similar observations have been made on melt crystallised samples by Hobbs (1971) for carbon fibres in polypropylene and by Burton and Folkes (1981) for carbon, glass and Kevlar fibres in nylon 6.6. The uniformity of the transcrystallinity along a fibre appears to be related to the type of fibre and also perhaps to the presence of a size coating. For example, Fig. 3.14

shows that Kevlar fibres are so effective at nucleating spherulitic growth, that a virtually uniform sheath of transcrystalline material surrounds the fibre. The material composing this sheath is anisotropic, as judged from the systematic extinction observed when the fibre is rotated between cross-polars in the optical microscope. In the case of carbon fibres, the transcrystallinity is more variable compared to Kevlar, although the greatest differences are found with type of fibre e.g. high modulus or high strength rather than with surface treatment. Glass seems far less effective in nucleating spherulitic growth.

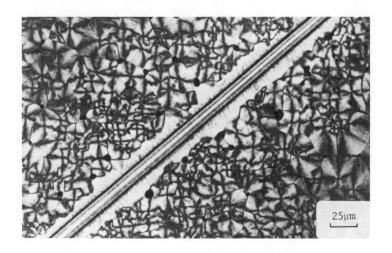

FIG. 3.14. Optical micrograph showing transcrystallinity around a Kevlar fibre in nylon 66.
(After Burton and Folkes, 1981)

The presence of such a significant proportion of transcrystalline material around a fibre should have an effect on composite physical properties. Bessell and Shortall (1975) have shown from composite fracture surface studies, that when fibre pull-out occurs it is possible for the fibres to be surrounded by sheaths of the matrix material. They suggest that this may be related to the presence of transcrystalline material, which results in a weak interface between the columnar growth structure adjacent to the fibres and main spherulitic structure in the matrix. Unexpectedly large ℓ_c values have been found from fracture surface studies by Blumentritt et al (1975b) and Burton and Folkes (1981) using nominally well-bonded fibre-matrix combinations. However, at the volume fractions of fibres used in these experiments, the columnar crystalline regions around adjacent fibres interleave and thereby modify the matrix material as a whole. It is not obvious, therefore, at the present time whether the observed features of the fracture surface originate from the mechanism proposed by Bessell and Shortall (1975) or are a direct consequence of the presence of anisotropic material surrounding the fibre. Nevertheless, irrespective of the detailed mechanisms involved, the effect of fibre-matrix interactions on the interfacial shear strength is not insignificant and is clearly relevant to the whole subject of optimisation of fibre surface treatments. These interactions also manifest themselves in composites based on non-crystalline thermoplastics. Thus Kardos (1973) has reported dramatic increases in the strength of poly-carbonate - carbon fibre composites, following an annealing procedure applied after composite fabrication - an effect attributed to a modification of the fibre-matrix interfacial morphology.

References

Anderson, R.M. and Lavengood,R.E. (1968). Variables affecting strength and modulus of short fiber composites. S.P.E. Journal, 24, 20-26.

Arridge, R.G.C. (1975). The effect of interlayers on the transverse stresses in fibre composites. Polym. Eng. Sci., 15, 757-760.

Bascom, W.D., Cottington, R.L. and Timmons, C.O. (1977). Fracture reliability of structural adhesives. J. Appl. Polym. Sci. : Applied Polymer Symposium, 32, 165-188.

Bessell, T. and Shortall, J.B. (1975). The crystallization and interfacial bond strength of nylon 6 at carbon and glass fibre surfaces. J.Mat.Sci., 10, 2035-2043.

Blumentritt, B.F., Vu, B.T. and Cooper, S.L. (1974). The mechanical properties of oriented discontinuous fiber reinforced thermoplastics: I Unidirectional fibre orientation. Polym. Eng. Sci., 14, 633 - 640.

Blumentritt, B.F., Vu, B.T. and Cooper, S.L. (1975a). Mechanical properties of discontinuous fibre reinforced thermoplastics: II Random-in-plane fiber orientation. Polym. Eng. Sci., 15, 428-436.

Blumentritt, B.F., Vu, B.T. and Cooper, S.L. (1975b). Fracture in oriented short fibre reinforced thermoplastics. Composites, 105-114.

Brody, H. and Ward, I.M. (1971). Modulus of short carbon and glass fibre reinforced composites. Polym. Eng. Sci., 11, 139-151.

Broutman, L.K. and Agarwal, B.D. (1974). A theoretical study of the effect of an interfacial layer on the properties of composites. Polym. Eng. Sci., 14, 581-588.

Burton, R.H. and Folkes, M.J. (1981). To be submitted to Plastics and Rubber Processing and Applications.

Chen, P.E. (1971). Strength properties of discontinuous fiber composites. Polym. Eng. Sci., 11, 51-56.

Cox, H.L. (1952). The elasticity and strength of paper and other fibrous materials. Brit. J. Appl. Phys., 3, 72-79.

Curtis, P.T., Bader, M.G. and Bailey, J.E. (1978). The stiffness and strength of a polyamide thermoplastic reinforced with glass and carbon fibres. J. Mat. Sci., 13, 377-390.

Halpin, J.C. and Kardos, J.L. (1978). Strength of discontinuous reinforced composites: I Fibre reinforced composites. Polym. Eng. Sci, 18, 496-504.

Hobbs, S.Y. (1971). Row nucleation of isotactic polypropylene on graphite fibres. Nature, 234 , 12-13.

Kardos, J.L. (1973). Regulating the interface in graphite/ thermoplastic composites. J. Adhesion, 5, 119-138.

Kelly, A. and Tyson, W.R. (1965). Tensile properties of fibre reinforced metals: Copper/tungsten and copper/molybdenum. J. Mech. Phys. Solids, 13, 329 - 350.

Krenchel, H. (1964). Fibre Reinforcement. Academisk Forlag, Copenhagen.

Lavengood, R.E. (1972). Strength of short-fiber reinforced composites, Polym. Eng. Sci., 12, 48-52.

Lees, J.K. (1968a). A study of the tensile modulus of short fiber reinforced plastics. Polym. Eng. Sci., 8, 186-194.

Lees, J.K. (1968b). A study of the tensile strength of short fiber reinforced plastics. Polym. Eng. Sci., 8, 195-201.

Marom, G.and Arridge, R.G.C. (1976). Stress concentrations and transverse modes of failure in composites with a soft fibre-matrix interlayer. Mat. Sci. Eng, 23, 23-32.

McNally, D. (1977). Short fiber orientation and its effect on the properties of thermoplastic composite materials. Polym. - Plast. Technol. Eng., 8(2), 101-154.

Masoumy, E., Kacir, L. and Kardos, J.L. (1980). Effect of fiber aspect ratio and orientation on the stress-strain behaviour of aligned, short fiber reinforced, ductile epoxy. Report,   cont...

58

Materials Research Laboratory, Washington University, St Louis, Missouri.

Ogorkiewicz, R.M. (1971). Mechanical properties of reinforced thermoplastics. Composites, 2, 29-32.

Peiffer, D.G. (1979). Impact strength of thick-interlayer composites. J. Appl. Polym. Sci., 24, 1451-1455.

Phillips, D.C. and Harris, B. (1977). The strength, toughness and fatigue properties of polymer composites. Polymer Engineering Composites, chapter 2, ed. M.O.W. Richardson. Applied Science Publishers, Barking, Essex.

Piggott, M.R. (1966). A thory of fibre strengthening. Acta Met., 14, 1429-1436.

Ramsteiner, F. and Theysohn, R. (April 1979). Tensile and impact strengths of unidirectional, short fibre-reinforced thermoplastics. Composites, 111-119.

Tsai, S.W. and Pagano, N.J. (1968). Composite Materials Workshop, eds. Tsai, Halpin and Pagano. Technomic Publishing Co, Stamford, Conn.

Weiss, R.A. (1980). Mechanical properties of polypropylene reinforced with short graphite fibers. Submitted to Polym. Eng. Sci.

Whitney, J.M. and Riley, M.B. (1966). Elastic properties of fiber reinforced composite materials. Am. Inst. Aeronaut. Astronaut.J., 4, 1537 - 1542.

CHAPTER 4
Compounding

The term "compounding", in the context of this book, refers to the
methods that are used to introduce the fibres into the polymer
matrix prior to the moulding of a component. It has gradually become
apparent that the detailed procedures for compounding have a marked
effect on the properties of the moulded component and a significant
amount of work is now being directed towards a more fundamental
appreciation of the processes involved. For companies who market
fibre filled thermoplastics, the understanding and close control
of the compounding process is obviously of paramount importance.
In most cases therefore such information is proprietory or is
protected by patents.

The ultimate criterion of success for any compounding method is
whether the final moulded component has the desired properties, such
as high stiffness and good surface finish. For this to be achieved
the following requirements normally have to be met:-

(a) The fibres have to be "wetted-out" i.e. each fibre must be
totally enclosed by the matrix.

(b) The fibres should be uniformly dispersed throughout the matrix,
with an absence of undispersed fibre bundles, which might otherwise
lead to variable strength of the composite.

(c) The fibres should be of sufficient length compared to their
diameter to ensure an effective transfer of stress from the matrix
to the fibres.

At first sight, it would seem that there could be little problem in meeting these requirements by a judicious control of the compounding conditions. However for many polymer-fibre combinations it is remarkably difficult to arrive at this situation, using commercially viable methods. For example, attempts to obtain good fibre dispersion lead only too often to unacceptably low fibre lengths.

Production of components in short fibre reinforced thermoplastics is usually carried out by injection moulding. In principle, therefore, if a screw injection moulding machine is used, the compounding and component fabrication stages can be combined in one operation i.e. the feedstock could be a dry blend of the polymer and fibres in the appropriate proportions. In this case, one is relying primarily on the screw-back stage in effecting a good dispersion of the polymer and fibres. However, attempts to injection mould dry blends directly can lead to problems of poor surface finish and variable strength, due to the presence of undispersed fibre bundles in the finished article. It is more usual, therefore, to compound the dry blend separately to produce pelletized feedstock, which is subsequently injection moulded. For the majority of moulding companies, the use of the pelletized compound is much more convenient, eliminates the need for special materials handling equipment and minimises the health hazards associated with airborne fibres. Although the use of a separate compounding operation produces a feedstock which can be moulded very satisfactorily, it does introduce another possible variable. Moulders are not only interested in achieving good properties of their mouldings but require that these are achieved consistently over a long period of time. Even in the case of unfilled polymers, unpredictable variations in polymerization and moulding conditions lead to significant scatter in physical property data. For short fibre reinforced thermoplastics, though, it is generally accepted that is is the unexpected variations in fibre length and dispersion that are the main contributions to the inconsistencies in product properties. The quantitative measurement of fibre length distribution therefore features prominently in compounding and injection moulding work especially in view of the dependence of composite properties on fibre length as discussed in Chapter 2. The variation of fibre length distribution with processing conditions is a reasonable measure of the relative importance of the various machine parameters. Having isolated some of the principal factors influencing fibre length, attempts can then be made to improve on the control of the relevant machine parameters, especially through the use of adaptive control technology.

4.1 ASSESSMENT OF FIBRE LENGTH AND DISPERSION

The fibres that are present in the pre-mix granules and in the final moulded component can have a very wide distribution of lengths. It is largely a matter of personal choice how one presents these data. One of the most popular measures of fibre length is the modal value, which is the length corresponding to the maximum in the distribution of fibre lengths. If more detailed information concerning the fibre length distribution is required, then number or weight average lengths can be quoted or in the extreme case the complete histogram of lengths can be given. Obtaining the latter is simple in principle, but considerable experimental care must be exercised if meaningful histograms are to be obtained in practice.

Fibre length distributions are best obtained by manual measurement of the lengths of a large number of fibres taken from a representative sample. Recently, with the advent of image analysis equipment e.g. Quantimet and Optomax, this very tedious procedure has been made considerably easier and more rapid. Nevertheless, the crucial part of the exercise is the sample preparation. Although techniques are available for directly viewing the fibres in sections cut from the bulk material, (see section 5.1) these are unsuitable for the unambiguous assessment of fibre length. This arises because these techniques only give projected images of the fibres in the viewing plane. It is necessary, therefore, to recover the fibres by selective removal of the matrix, so that these can be viewed when lying in a single plane. In this case, any representative portion of the composite is taken which may or may not be a thin section. The technique used for removal of the matrix obviously depends on the particular fibre/matrix combination, but generally falls into one of the following three categories:-

(i)   High temperature ashing by pyrolysis, involving temperatures well above the melting point of the matrix.

(ii) Low temperature ashing using plasma oxidation and conducted below the melting point of the matrix.

(iii) Chemical digestion using solvents or acids.

High temperature ashing is frequently used to retrieve the glass fibres in reinforced thermoplastics, but care must be taken to avoid fibre breakage due to thermal shock when the sample is cooling. Although this technique is simple to apply, a study of the recovered fibres using scanning electron microscopy reveals the presence of considerable debris remaining on the fibre surface, which inhibits complete separation of the fibres. Fig. 4.1, taken from the work of Sawyer (1979), illustrates this point rather well. This problem can be particularly serious if image analysis is to be utilised for the measurement of fibre length, since touching fibres are recorded as a single feature. In the case of glass fibres, the difficulty is made more acute by the use of the size which is applied to the fibres to promote coupling with the matrix. Even after low temperature ashing, when the organic materials in the size have been oxidised, the inorganic silanes remain and cause fibre-fibre adhesion. Sawyer (1979) has investigated this problem and has found that an aqueous dispersion of a monofunctional silane with a glass lubricant reacts with the silane on the glass surface to form methyl siloxanes, which promotes separation of the glass fibres. Image analysis can then be applied to these fibres.

Fibre length distributions in extrusion compounded material can be very wide, ranging from a few tens of microns to the length of the original fibres used as feedstock in the compounding operation, which is often about 5mm. To reduce further the instances of crossed or touching fibres when measuring their length, a sieve stack may be used to separate the fibres into samples having much narrower fibre length distributions. This approach has been adopted by Lunt

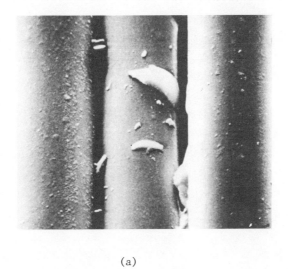

(a)

5μm

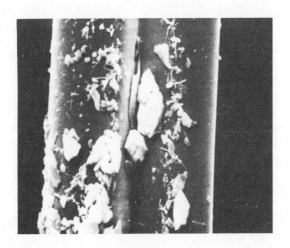

(b)

FIG. 4.1. Scanning electron micrographs of (a) original short glass
fibres before compounding with the matrix resin (b) glass fibres
recovered from the composite pellets using high temperature ashing.
(After Sawyer, 1979).

and Shortall (1979) in a study of the compounding of glass fibres in nylon 66. Here they used eight sieves having a mesh size ranging from 0.045 - 4.0 mm. The fibre length distributions for each sieve could be obtained in the normal manner using a Ziess particle size analyser - see Endter and Gebauer (1956). By means of a computer, the individual sieve distributions can be combined to give the fibre length distribution for the whole sample. This method provides a very accurate means of determining distributions but is time consuming. Nevertheless, it provides the means whereby accuracy can be achieved without the need to invest in additional modules for image analysis equipment.

For certain composites e.g. carbon fibre reinforced thermoplastics, high temperature ashing cannot be used to retrieve the fibres since they suffer major degradation. In cases such as this, low temperature ashing or chemical digestion must be used, the latter being more widely applied at the present time. The reader is referred to manufacturers data for information regarding appropriate solvents.

4.2   COMPOUNDING TECHNIQUES

4.2.1 Introduction

Aside from the direct injection moulding of a dry blend of fibres and polymer, a number of different techniques exist for the production of pre-mix pellets for injection moulding. The selection of the appropriate compounding technology is very dependent on the requirements of fibre length, volume fraction and degree of dispersion of the fibres throughout the matrix. One of the most common methods of compounding involves the use of a single or twin screw extruder to mix chopped fibres and matrix and produce an extrudate, which can be pelletized to give roughly spherical granules. This technique can cause very significant fibre breakage but nevertheless produces well dispersed fibres. As discussed later, the conditions used during compounding have a pronounced effect on fibre length and dispersion. However, there are fundamental limitations with this technique if pre-mix pellets are required having a high volume fraction of long fibres. The fact that it is difficult, if not impossible, using normal extrusion compounding techniques to meet this requirement is due in part to the excessive fibre breakage occurring along the screw, but also to the difficulties of feeding the dry blend of fibres and matrix, when large concentrations of fibres are involved. These two factors are not entirely attributable to limitations of the processing equipment but are related to the much more fundamental fact that the maximum possible packing of randomly oriented fibres is very low – Parratt (1972), Charrier (1975). Attempts can be made to increase the volume fraction of fibres beyond their natural packing density either by application of pressure (as will occur in an extruder) or by feeding the extruder with a higher volume fraction of fibres, as is commonly carried out commercially. Under these circumstances, for the fibres to conform to packing requirements, they must either break into shorter fragments or form bundles within which the fibres are fully aligned. The compressibility of different fibre materials has been reported by Krenchel (1964), who has shown that even modest pressures ($\sim$0.5 – 1.0 MPa) applied to a non-woven glass fabric results in fibre breakage. This problem will be even more pronounced in the case of carbon fibres.

To produce pellets having a large volume fraction of long fibres, a completely different approach is required. Most of the techniques developed rely on the fact that efficient fibre packing can only be achieved when the individual fibres are aligned parallel to each other. The theoretical maximum fibre volume fraction in this case is very high ($\sim$ 90%). The usual approach to the production of such pellets is by the continuous coating of fibre rovings with the matrix polymer, followed by pelletizing. This method produces pellets having highly aligned fibres with a length approximately equal to the pellet. Possible disadvantages of this technique are that the fibres may not be adequately dispersed and there may be imperfect wetting of the fibres by the matrix, especially for those fibres near the centre of the pellet.

Pellets having well collimated fibres, completely wetted in the matrix, can be prepared by the in situ polymerization of the

thermoplastic matrix around the individual fibres - Baer (1975).
However, it must be remembered that the production of pre-mix pellets
is carried out to a large extent to simplify the materials handling
procedures, during subsequent moulding. The moulding stage itself
results in improvements in fibre dispersion and "wet-out" but fibre
breakage will also occur. As pointed out by Bader and Bowyer (1973),
there is little advantage in developing improved compounding
techniques to retain fibre length, unless modifications are also
made to the injection moulding procedures. The latter has met with
moderate success but there is still considerable scope for the
development of novel injection moulding techniques, aimed at the
retention of fibre length.

The widespread use of compounding procedures followed by the injection
moulding of pre-mix pellets is based on the accumulated evidence
that attempts to mould fibre/matrix blends directly can lead to
problems of poor surface finish on the moulding and inferior
properties - see section 4.3. Although in principle, it should be
possible to compound and mould satisfactorily in one process, it
may be that the requirements to conserve fibre length yet have good
fibre "wet-out" are mutually exclusive without resource to a separate
compounding stage.

4.2.2 Extrusion Compounding

Melt compounding of fibres and polymer can be carried out using either
single or twin screw extruders. In cases where bulky fillers and
reinforcements e.g. asbestos are to be incorporated, it is sometimes
preferable to use internal mixers e.g. a Banbury. Generally however,
extrusion compounding is the most widely used method commercially.

Compounding extruders must incorporate special design features to
cope with the abrasive nature of some fibres, especially glass, the
high powers required for compounding and possibly novel methods for
feeding the polymer and fibre. The machine wear, especially on the
screw and barrel, can be quite unacceptable unless special nitrided
steels or hard alloy coatings are used. The screw design is of
paramount importance and since one of the major sources of fibre
breakage occurs at the feed section, it is usual to have deep flights

to minimize this. A vented barrel is often used, since the surface
coatings on some fibres can partially decompose on heating. The die
design is especially important in single screw extruders, since the
production of a homogeneous melt relies on the development of a high
die head pressure. In the case of a twin screw machine the mixing
is effected almost entirely by the action of the screws.

It is in the versatility of the feeding arrangements of the polymer
and fibre that is one of the distinguishing features of the twin
screw compounding extruder. In both the single and twin screw
machine, the feeding arrangement may be of the conventional form i.e.
a dry mix of the polymer and fibre is fed into the hopper. It is
also possible to introduce the fibres as a continuous strand at the
hopper and rely on the screw to break the fibres into short lengths.
However, there is the possibility when using a twin screw machine,
to introduce the fibres (either continuous strand or chopped) at some
point along the barrel where the polymer is already fully molten.
The twin screw action readily takes up the fibre - a feature that is
not normally exhibited by a single screw machine, unless special
consideration is given to the design of the feed port. The addition
of fibres to a pre-melted polymer has the advantage that less fibre
breakage occurs together with an improvement in dispersion. There is
the additional advantage that much less wear takes place in the
working parts of the extruder.

Apart from the single and twin screw extruder, a number of other
specially designed compounding machines are available for the
production of reinforced thermoplastics. One such machine, named
the Buss Ko-Kneader, derives its compounding action from the rotation
and reciprocation of a screw, with partly discontinuous flights, in
a toothed barrel. In this case, the compounded material is extruded
at a non-uniform rate, and so it is frequently used as feedstock for
a conventional single screw extruder.

For any particular extruder, the processing variables have a
significant effect on the extent of fibre breakage, dispersion and
fibre "wet-out". Preliminary studies usually concentrate on an
assessment of the fibre length. This is because although the

particular compounding conditions may lead to excellent wetting of the matrix and fibres, the fibres may have sub-critical lengths and so will be of little use for reinforcement. A more comprehensive assessment must quantify the degree of bonding at the fibre/matrix interface. A generalised approach to this problem has been reported by a number of workers - see for example Bader and Bowyer (1973), McNally et al (1978) and Lunt and Shortall (1979). This will be discussed in section 4.3.

DIE DIAMETER mm	SCREW SPEED r.p.m.	WEIGHT AV. LENGTH mm	NUMBER AV. LENGTH mm	REINFORCEMENT EFFICIENCY f
1.5	20	1.05	0.47	0.87
	40	0.71	0.42	0.83
	80	0.39	0.31	0.76
2.0	20	1.11	0.54	0.88
	40	0.85	0.43	0.84
	100	0.65	0.34	0.78
3.0	20	1.43	0.66	0.91
	40	0.59	0.36	0.80
	100	0.57	0.35	0.79
4.0	20	1.96	0.92	0.94
	40	0.99	0.47	0.86
	100	0.69	0.40	0.82

TABLE 4. Experimental dependence of the reinforcment efficiency factor (f) on die diameter and screw speed for glass fibre reinforced nylon 66. Compounding conducted using a Betol 25 mm single screw extruder.

$$f = \frac{\text{Composite strength calculated from measured fibre length distr.}}{\text{Composite strength calculated using the initial fibre length}}$$

(After Lunt and Shortall, 1979)

Lunt and Shortall (1979), in a study of the compounding of glass fibres in nylon 6.6, have found a marked dependence of fibre length on screw speed and die diameter, using a Betol 25mm single screw extruder. Table 4, taken from their paper, shows that for each die size, increasing screw speed causes an increase in fibre breakage.

The dependence of fibre breakage on die diameter for a given screw
speed is not straightforward. The fibres were nominally 3mm long
at the commencement of compounding. Modest changes in barrel
temperature led to significant changes in fibre length e.g. an
increase in temperature from $255^{o}C$ to $290^{o}C$ increased the number
average length by about 25%. Similar observations have been made
by Folkes and Stuart (1979) in the compounding of carbon fibres
into nylon 6.6, but in this case, the increase in fibre length for
a given temperature rise was much larger. This may be related to
the relative brittleness of carbon as compared to glass fibres.
The dependence of fibre breakage on the volume fraction of glass is
particularly interesting. As shown in Table 5, again taken from the
work of Lunt and Shortall (1979), there is an initial decrease in
fibre length as the glass concentration increases. For weight
fractions greater than 30% glass, the fibre length increases as the
concentration increases, but the fibres are less well disperesed in
the matrix. At low fibre concentrations, the fibres are well
separated from each other and so fibre-fibre breakage will be
minimal. On the other hand at very high fibre concentrations, the
rheology of the composite is significantly modified in such a way
that during flow, a large proportion of the material is immobile.
This factor would again tend to reduce fibre-fibre breakage.

Locating the main sources of fibre breakage during extrusion
compounding requires a systematic examination of the fibre length
distribution in samples removed from the various stages in the
process. Very little work of this type has been reported. Kashfi
(1976) has examined the fibre length distribution in glass fibre
filled polypropylene at the various points along the screw in a
single screw extruder. In this case, the extruder was operated
until steady state conditions had been reached. The process was
then stopped and the screw extracted. The distribution of the
polymer in the screw flights was then apparent, as shown in Fig. 4.2.
Samples could be removed from the flights and at other points in
the extruder e.g. the die entry point. These data show that
substantial breakage, ~20%, occurs at the rear end of the screw,
hence confirming other less direct observations. Another significant

GLASS CONTENT wt %	WEIGHT AV. LENGTH mm	NUMBER AV. LENGTH mm	REINFORCEMENT EFFICIENCY f
5	1.19	0.65	0.90
10	0.72	0.41	0.83
30	0.71	0.40	0.83
50	1.23	0.50	0.87
60	1.76	0.75	0.92

TABLE 5. Experimental dependence of the reinforcement efficiency
  factor (f) on glass content for glass fibre reinforced nylon 66.
  Compounding conducted using a Betol 25mm single screw extruder.

$$f = \frac{\text{Composite strength calculated from measured fibre length distr.}}{\text{Composite strength calculated using the initial fibre length}}$$

  (After Lunt and Shortall, 1979)

source of breakage occurs at the die entrance, where fibres are
forced to rotate from a previously misaligned state into the flow
direction. This is also found to be the case during capillary
rheometry - Bell (1969). More recently, Lunt and Shortall (1980)
concluded that the major source of fibre breakage during extrusion
compounding of glass fibre reinforced nylon 66 occurs in the melting
region of the screw.

4.2.3 Wire coating techniques

With the aim of producing a moulding pellet containing a high
volume fraction of long fibres, an alternative method to extrusion
compounding is required. One method utilises wire coating technology
and consists of passing a bundle of continuous rovings or tows of
fibres through a specially designed extruder die-head, so that the
tows are coated and partially impregnated by the molten polymer.
The impregnated fibre tow is chilled in a water bath and then chopped

(a)

(b)

FIG. 4.2. Distribution of polymer along the flights of an extruder
screw; glass fibre reinforced polypropylene extruded using a
screw speed of (a) 60 r.p.m. (b) 20 r.p.m.
(After Kashfi, 1976)

into pellets of the desired length. An experimental arrangement
which utilises a cross-head die is shown in Fig. 4.3 and is taken
from the work of Bader and coworkers (1973). An additional advantage

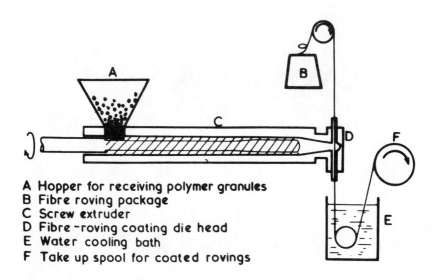

A Hopper for receiving polymer granules
B Fibre roving package
C Screw extruder
D Fibre-roving coating die head
E Water cooling bath
F Take up spool for coated rovings

FIG. 4.3. Equipment for the production of continuous polymer coated
fibre-rovings. The final pelletizing stage is not shown.
(After Bader and Bowyer, 1973)

of this approach is that a conventional extruder can be used, since
it is no longer required to process a fibre/polymer mixture. The
technique has been applied successfully to the production of
compounds containing over 40% v/v glass or 30% v/v carbon. The
Fiberfil Division of Dart Industries Inc., USA, market moulding
compounds of this type with glass fibres as the reinforcement.
A difficulty that can arise with this technique is the uneven

penetration of the fibre roving by the viscous polymer. This may result in a concentration gradient of polymer decreasing in the direction of the fibre roving core, which may remain incompletely wetted and impregnated by the polymer. However, further dispersion of the fibres will occur during subsequent moulding. Cloud and Schulz (1975) point out that moulding pellets based on this compounding technique can result in excessive machine wear, since in order to achieve good dispersion, considerable mechanical work has to be done in the moulding machine.

An additional advantage of this compounding method is apparent in the manufacture of short-fibre hybrid composites. Here, a short fibre filled thermoplastic can be used to coat continuous rovings of another fibre type. After pelletising in the usual way, the two fibre species will be partially mixed but one of the species will have a well defined length in the composite. If the continuous rovings were carbon, then this fibre can be used primarily for composite stiffening while the other fibre species, e.g. glass, that is precompounded in the thermoplastic, can be used to increase the work of fracture of the composite.

4.2.4 In-situ polymerization

Baer (1975) has described a novel method of producing pellets, having a large volume fraction of short fibres, which are thoroughly and uniformly wetted by the matrix. The work reported confines itself with the encapsulation and collimation of short glass strands in styrene-acrylonitrile (SAN), but previous work by Andersen and Morris (1968) describes techniques applicable to boron or glass fibres in thermosetting epoxy resins. The process described by Baer involves polymerising styrene and acrylonitrile in the presence of loose (~ 5mm long) strands in an aqueous suspension system. The reactors are charged with glass strands and after removal of air and nitrogen purging, the SAN monomers are introduced. After a short time has elapsed, to allow thorough wetting of the strands by the monomer, an aqueous solution of a protective colloid is introduced. The reactors are sealed and polymerisation proceeds. By careful control of the polymerisation conditions, especially the composition and concentration

of protective colloid, the strands align themselves in a parallel
array completely embedded by the polymerised matrix, as shown
in Fig. 4.4. The mechanism of collimation is thought to be initiated
by absorption of monomer by the glass strands through capillary
action, occurring between the individual glass fibres composing the
strands. Surface energy forces attract the strands and prevent their
separation. Stacks of strands thus form, as shown diagrammatically
in Fig. 4.5. Colli ation of the strands is encouraged by gentle
agitation of the aqueous medium, which results in sliding of the
strands to form fully collimated structures.

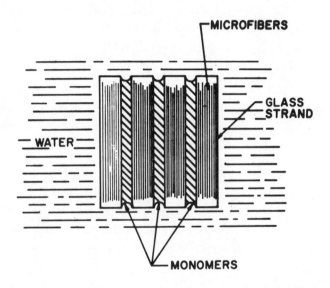

FIG. 4.5. Schematic representation of the collimation of glass
strands, prewetted by monomer, and suspended in an aqueous medium.
(After Baer, 1975)

Injection mouldings prepared from these pellets are found to have
superior properties compared to mouldings produced from commercial
blends or dry blends of glass fibre and SAN. Furthermore, these
properties, especially the tensile modulus and strength, are relatively

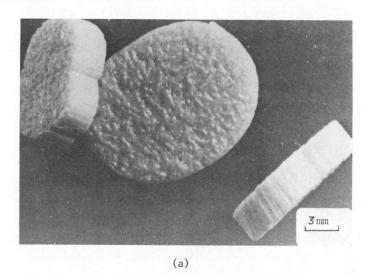

(a)

(b)

FIG. 4.4. Encapsulation and collimation of glass strands in styrene acrylonitrile (a) 66% wt of $\frac{1}{8}$" strands (b) 71% wt of $\frac{1}{4}$" strands. (After Baer, 1975).

insensitive to the injection moulding conditions. This is, in part, attributed to the greater protection afforded to the glass fibres by virtue of their coating of polymer. This seems to be consistent with similar observations made with pellets produced using the wire coating technique, although this does not result in such an effective wetting of the fibres, as in the case of in-situ polymerisation. The technique should be applicable to other polymers by judicious choice of the polymerisation conditions, but to date no such work has been reported.

4.3. GENERAL ASSESSMENT OF COMPOUNDING TECHNIQUES

As seen through the whole of this chapter, the various compounding methods available have a major influence on both fibre length (and dispersion) and the effectiveness with which the matrix coats the fibres during the compounding process. The latter will significantly affect the fibre/matrix interfacial shear strength and should be evaluated, together with the fibre length when assessing the efficiency of the various compounding methods. One approach that has been adopted by McNally et al (1978) and Lunt and Shortall (1979) utilises the Kelly-Tyson elastic fibre/plastic matrix model, described in Chapter 2. For the case of misaligned short fibres in a plastic matrix the theoretical strength of the composite can be expressed by

$$\sigma_{uc} = K \left\{ \sum_{\ell_i = 0}^{\ell_i = \ell_c} \frac{v_i \tau_u \ell_i}{d} + \sum_{\ell_j = \ell_c}^{\ell_j = \infty} \sigma_{uf} v_j \left(1 - \frac{\ell_c}{2\ell_j}\right) \right\} + (1-v)\sigma_m'$$

where the first term in the curly brackets is the contribution of the sub-critical fibres and the second term that of the super-critical fibres. The final term is the matrix contribution. The coefficient K is an orientation factor ($\leqslant 1$).

Lunt and Shortall (1979) in their work on compounding of glass fibres in nylon 66, assume a value for $\tau_u = 60$ MNm^{-2} and value for K = 0.825 (the reason for choosing this particular value is not obvious). Since the other parameters in the equation are known, a measurement of the fibre length distribution enables the composite strength to be calculated. If this quantity is divided by the value of composite strength that would have been achieved if the fibres remained at their

initial length (keeping all other quantities fixed) then a value for
the fibre reinforcement efficiency f, can be obtained. The authors
consider that this is a more meaningful indication of fibre
degradation occurring during compounding than would be achieved by
a fibre distribution measurement alone, since it allows for the non-
linear dependence of composite stress on fibre length. Table 4,
taken from the work of Lunt and Shortall (1979) illustrates the
variation of f with screw speed and die diameter, using a single
screw extruder. Evidence for the additional fibre breakage occurring
during injection moulding is shown in Table 6, containing data
extraced from their work. In this case, the compounded material
was moulded on a Stubbe SKM 76 injection moulding machine to produce
ASTM tensile bars.

COMPOUNDING CONDITIONS		MEASURED TENSILE STRENGTH (DRY) MNm^{-2}	MEASURED TENSILE STRENGTH (WET) MNm^{-2}	CALCULATED TENSILE STRENGTH (WET) MNm^{-2}	CALCULATED TENSILE STRENGTH (DRY) MNm^{-2}	REINFORCEMENT EFFICIENCY f' (DRY)	REINFORCEMENT EFFICIENCY f' (WET)
DIE DIAMETER mm	SCREW SPEED rpm						
1.5	40	171	76	219	134	0.78	0.57
2.0	40	165	77	219	134	0.75	0.58
3.0	40	179	83	199	113	0.90	0.74
4.0	40	187	80	219	130	0.85	0.62
4.0	60	187	85	219	129	0.87	0.66
4.0	90	189	80	210	119	0.90	0.67

TABLE 6. Fibre reinforcement efficiency factors (f') for ASTM tensile
bars of short glass fibre reinforced nylon 66. The original
compounding was conducted using a Betol 25mm single screw extruder.
The tensile bars were moulded on a Stubbe SKM76 injection
moulding machine.

$$f' = \frac{\text{Measured tensile strength of the bars}}{\text{Tensile strength calculated from measured fibre length distr.}}$$

(After Lunt and Shortall, 1979)

The approach adopted by McNally et al (1978) is rather different
in detail from that of Lunt and Shortall (1979). The Kelly-Tyson
model can be used to describe both the composite strength and the

elastic modulus; in both cases the form of the expression for σ_{uc} and E_o is similar - see Chapter 2. Now σ_{uc} depends on K, τ_u and the fibre length distribution (together with other quantities which we assume constant). Likewise E_o depends on an orientation factor K', τ_u and the fibre length distribution. On the basis of strength and modulus anisotropy data, McNally et al (1978) suggest that K and K' have very similar values. Using the measured values of σ_{uc} and E_o, together with the fibre length distribution and an assumed value for τ_u, enables values for K and K' to be computed. In general this will yield different values for these two quantities. The process of iteration is continued using different values of τ_u, until the computed values of K and K' converge. The corresponding value of τ_u can be substantially less than the shear yield strength of the matrix, indicating imperfect load transfer from the matrix to the fibres during composite loading. The utility of this analytical technique has been demonstrated by comparing the results for τ_u using various compounding methods. Some results taken from the work of McNally et al (1978) are shown in Table 7, and serve as an excellent example of the very large variation in τ_u obtainable, using different compounding methods.

The same approach can be used for examining the effects on τ_u of the age of the compounding extruder, the screw configuration and the conditions used during the subsequent injection moulding of the compounded material.

MATERIAL	COMPOUNDER	NUMBER AVERAGE LENGTH mm	WEIGHT AVERAGE LENGTH mm	INTERFACIAL SHEAR STRENGTH τ_u MNm^{-2}
ACETAL COPOLYMER	W	0.15	0.36	6.07
25% wt. GLASS FIBRES	Z	0.14	0.32	8.07
POOR FIBRE -	Y	0.21	0.46	4.41
MATRIX BONDING	X	-	-	-
	DIRECT INJECTION	0.28	1.14	3.65
ACETAL COPOLYMER	W	0.20	0.43	34.06
25% wt. GLASS FIBRES	Z	0.18	0.36	37.06
GOOD FIBRE -	Y	-	-	-
MATRIX BONDING	X	0.28	0.52	19.65
	DIRECT INJECTION	-	-	-
POLYBUTYLENE TEREPHTHALATE	W	0.29	0.40	26.75
	Z	0.14	0.31	35.23
30% wt. GLASS FIBRES	Y	0.24	0.45	30.89
	X	-	-	-
	DIRECT INJECTION	0.34	0.93	12.69

TABLE 7. Dependence of the interfacial shear strength (τ_u) on
compounding method for short glass fibre reinforced thermoplastics.
The tensile bars were moulded using either a Stubbe or Stokes
injection moulding machine. The letters W, X, Y and Z refer to
the different compounding methods employed in this study but
their identity is not revealed in the reference source.

(Data extracted from McNally et al, 1978)

References

Andersen, H.M. and Morris, D.C. (1968) 23rd Annual Techn. SPI
    Conference, paper 17E.

Bader, M.G. and Bowyer, W.H. (July 1973). An improved method of
    production for high strength fibre reinforced thermoplastics.
    Composites, 150-156.

Baer, M. (1975). Composites obtained by encapsulation and collimation
    of glass fibers within a thermoplastic matrix by means of
    polymerization. J.Appl.Polym.Sci., 19, 1323-1336.

Bell, J.P. (1969). Flow orientation of short fibre composites.
    J.Comp.Mat., 3, 244-253.

Charrier, J.M. (1975). Basic aspects of structure-property
    relationships for composites. Polym.Eng.Sci., 15, 731-746.

Cloud, P.J. and Schulz, R.E. (1975). A primer on reinforcing and
    filling IM plastics. Plastics World, Sept. 22.

Endter, F. and Gebauer, H. (1956). A simple apparatus for the
    statistical evaluation of photographs in microscopy or electron
    microscopy. Optik, 13, 97-101.

Folkes, M.J. and Stuart, T. (1979). Unpublished work.

Kashfi, K. (1976). Extrusion of glass fibre filled polypropylene.
    M.Sc. Thesis, Department of Materials, Cranfield Institute of
    Technology, Bedford.

Krenchel, H. (1964). Fibre Reinforcement. Academisk Forlag,
    Copenhagen.

Lunt, J.M. and Shortall, J.B. (Sept. 1979). The effect of extrusion
    compounding on fibre degradation and strength properties in short
    glass - fibre-reinforced nylon 6.6. Plastics and Rubber:
    Processing, 108-114.

Lunt, J.M. and Shortall, J.B. (June 1980). Extrusion compounding of
    short glass fibre filled nylon 6.6 blends. Plastics and Rubber:
    Processing, 37-44.

McNally, D., Freed, W.T., Shaner, J.R. and Sell, J.W. (1978). A method to evaluate the effect of compounding technology on the stress transfer interface in short fibre reinforced thermoplastics. Polym. Eng. Sci, 18, 396-403.

Parratt, N.J. (1972). Fibre-Reinforced Materials Technology. Van Nostrand Reinhold, London.

Sawyer, L.C. (1979). Determination of fiberglass lengths; sample preparation and automatic image analysis. Polym.Eng.Sci., 19, 377-382.

CHAPTER 5
Processing – Microstructure Correlations

The principal method used for the production of components in short
fibre reinforced thermoplastics is injection moulding. The normal
moulding cycle that is used for unfilled thermoplastics is also used
for the reinforced material but the detailed processing conditions
employed may be rather different. Since the properties of a short
fibre reinforced thermoplastic are very dependent on fibre length
and orientation, it is important that both of these parameters can
be controlled in the final moulding, by an appropriate choice of
processing conditions. The problem is aggravated by the fact that
when fibres are introduced into a thermoplastic, the rheological
properties of the melt are significantly modified. Furthermore, the
thermal conductivity of the melt is usually increased by the presence
of the fibres. Hence, both the flow field and thermal conditions
will be quite different compared to an unfilled thermoplastic. Much
experience has been gained with these materials by commercial moulders
and empirical procedures for obtaining good quality mouldings have
been developed - see e.g.Theberge (1973), Maxwell (1964), Lucius (1973),
Murphy (1965) and Hunter (1975). In addition, systematic research
using factorial analysis has been carried out, showing the extent to
which mechanical properties can be affected by processing conditions -
Schlich et al (1968), Bright (1980). During the moulding process,
the fibres may become oriented in a complex manner and confer on the
moulding a marked anisotropy of mechanical properties. The magnitude
of this anisotropy will depend on the fibre length and orientation
distribution.

The anisotropy may be calculated, using the procedures described in
Chapter 2, as well as measured experimentally. A valuable link may
therefore be formed between processing conditions and properties,
by a study of the microstructure of moulded components. The latter
will be discussed in this chapter and the manner in which micro-
structural information can be used to predict moulding properties
especially those relevant to design, will be the subject of Chapter 7.

During the moulding of fibre reinforced thermoplastics, special
processing conditions are recommended for the production of good
quality parts. These are listed below:-

(i)   High injection speed should be used in order to achieve a good
surface finish and to prevent premature solidification of the melt,
either in the cavity or at the gate.

(ii) The screw speed and back pressure must be kept to a minimum,
since although a homogeneous melt is required, fibre breakage may
become excessive.

(iii) The melt temperature used for reinforced thermoplastics is
usually at the upper end of the range recommended for the unfilled
counterpart. This is chosen to reduce the viscosity of the melt and
partly to assist in preventing premature solidification in the cavity.

(iv) After the cavity is filled, a long hold time is needed. This
is required, not only to ensure that the moulding dimensions are
correct, but to minimize the ever present problem of voiding observed
in the core of moulded components - Darlington and Smith (1975). This
is particularly important for reinforced thermoplastics, since the
shrinkage that must necessarily take place in the core of the moulding
at it is cooling down cannot be accommodated by "sinking" of the
surface layers, due to their intrinsic stiffness. The maximum hold
time is determined by the onset of gate freezing and of course by the
economic requirement of minimising the overall cycle time.

Certain other factors should also be taken into account for the
effective processing of reinforced thermoplastics. The wear and/or
corrosion of the screw and barrel may be greater if glass reinforced
grades are being moulded and so special alloys or hardened coatings

should be used wherever possible. Also, the design of the mould is
especially important for reinforced thermoplastics. There is the
obvious need to locate the gate(s) at realistic positions in order
to ensure the development of an appropriate fibre orientation
distribution in the final component. Although this is frequently
decided empirically, considerable guidance may be sought from a
knowledge of the fibre orientation developed in cavities having
simple geometries. The gate dimensions must also be larger than
would normally be appropriate for unfilled thermoplastics. This is
due to the greater likelihood of jetting occurring during the
filling of the cavity and of course the need to ensure that prolonged
hold times can be used, without gate freezing. In fact, for thick
section mouldings it is necessary to sprue gate or use pulsating
injection (Allan and Bevis 1981) otherwise unacceptable voiding
will occur in the core of the moulding. The reader who requires
guidance on the detailed aspects of injection mould design should
consult standard texts on this subject e.g. Pye (1968).

5.1 EXAMINATION OF MICROSTRUCTURE

Here we will be concerned with some of the techniques that can be
used to reveal the fibre orientation in reinforced thermoplastics,
together with some information concerning the nature of the bond
between the fibres and matrix.

5.1.1 Reflection microscopy of polished surfaces

This is one of the earliest techniques used to assess fibre
orientation in moulded components. A part of the moulding is
carefully polished down to 0.5μm alumina powder and the polished
section is observed in the optical microscope, using reflected light.
Fig. 5.1, taken from the work of Thomas and Meyer (1976), shows a
region in a tranverse section of a dumbbell test specimen. The
cylindrical fibres meet the polished surface in ellipses. The ratio
of the length of the major axis to minor axis will determine the
orientation of any chosen fibre, with respect to the surface:-

$\phi = \cos^{-1}$ (length of minor axis/length of major axis)
and ϕ is the angle between the fibre axis and normal to the viewing

surface. A further measure of fibre orientation is obtained by measuring the orientation of the major axis with respect to some reference axis, in the plane of the section. Although this method does provide some quantitative measure of fibre orientation, it is tedious to apply and only a small proportion of fibres are being sampled in any one section. Furthermore, the polishing process can cause quite serious damage to the fibres, as shown in Fig. 5.2., and this will also limit the accuracy of the orientation measurement.

5.1.2 Transmission optical microscopy

In thin mouldings of reinforced thermoplastics, containing less than about 10% volume fraction of fibres, a texture is often visible in transmitted light, which may be related to flow during the moulding process. An example of a short glass reinforced polypropylene is shown in Fig. 5.3. However Darlington et al (1976) stress the need for caution when interpreting results of this type, since the observable texture is confined to a central zone of the moulding. Conclusions regarding the possible anisotropy of mechanical properties would be very incorrect, if the texture alone was considered indicative of the fibre orientation throughout the whole moulding. Thus, in the case of the above moulding, stiffness measurements made on specimens cut parallel and perpendicular to the texture markings show that the stiffer direction is in fact perpendicular to these markings. The confusion caused by overlapping fibre orientation distributions can be minimized by cutting thin sections, which have a thickness such that only a few "layers" of fibres are being sampled. These sections may be obtained using microtomy or alternatively by using a diamond edged slitting saw. When the fibres are very different in colour from the matrix, they may be observed directly in transmitted light. An example of the ease with which carbon fibres may be observed is shown in Fig. 5.4. Glass fibres, on the other hand, are difficult to distinguish from the matrix and in this case radiography may be used very successfully, as described in the next section.

5.1.3 Radiography

The use of X-rays for observing inhomogeneities in materials e.g. cracks in metal components is well established in the field of Non-

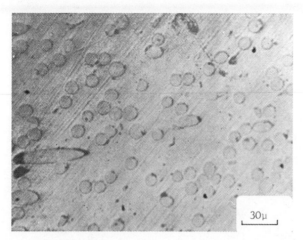

FIG. 5.1. Optical micrograph, taken in reflected light, of a
polished transverse section of an ASTM testbar, moulded in glass
fibre reinforced polytetramethyleneterephthalate.
(After Thomas and Meyer, 1976)

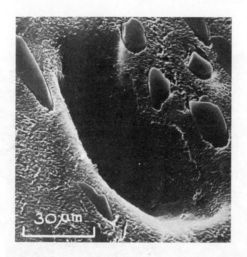

FIG. 5.2. Scanning electron micrograph of a polished section taken
from a glass fibre reinforced polypropylene injection moulded
disc, showing the presence of fibre damage.
(After Darlington and Smith, 1975)

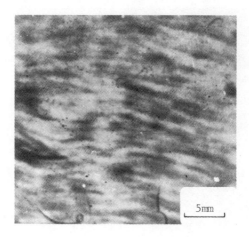

FIG. 5.3. Appearance of a region in a glass fibre reinforced
polypropylene plaque when viewed in transmitted light, showing
the presence of a texture.
(After Darlington et al, 1976)

FIG. 5.4. Transmission optical micrograph showing carbon fibres
in a reinforced nylon 66 moulding.

Destructive Testing. The technique relies on a variation of X-ray
absorption from one part of the sample to another and so, in principle,
could be applied to fibre reinforced thermoplastics. It is
unsuccessful, and indeed unnecessary, for carbon fibres in most
matrices, since both ingredients of the composite are carbon based.
However, it is a first class method for glass fibre reinforced
thermoplastics - Nemet et al (1962), Darlington et al (1976), Thomas
and Meyer (1976). There are two methods of approach that may be used.
One is referred to as macro-radiography and the other micro-radiography. In the former, the moulding itself is placed in contact with
a photographic plate and then exposed to a beam of X-rays. A typical
macro-radiograph taken from a disc moulding is shown in Fig. 5.5.
It has been confirmed that the observed texture is primarily
associated with fibre clumps and that no really useful information is
obtained on the well-dispersed fibres. To improve the resolution
of the technique, it is necessary to use thin sections (50-150µm) cut
from the moulding and a photographic plate of sufficiently high
resolution, to enable magnification of x 500 to be used. This is
referred to as micro-radiography and was orginally developed for use
with biological materials - Engstrom (1956). A contact microradiograph obtained from a thin section of glass fibre reinforced
polypropylene is shown in Fig. 5.6. The contrast between the fibres
and matrix is excellent and a much better assessment of the fibre
orientation distribution is possible compared to that using
metallographic polishing. The technique is also capable of identifying
the distribution of pigment in pigmented short fibre reinforced
thermoplastics - see Fig. 5.7. Folkes and Sharp (1981) have shown
how a thin section from a short fibre hybrid composite, containing
both glass and carbon fibres may be used to assess the separate
orientation distributions of the two fibre species, by combining
optical microscopy and X-ray contact microradiography - see Fig. 5.8.

5.1.4 Ultrasonic measurements

One method of detecting and obtaining a measure of structural
variations in short fibre reinforced thermoplastics is to examine the
propagation of ultrasonic waves through the material. Although this
does not provide a method for directly viewing the fibres, it is

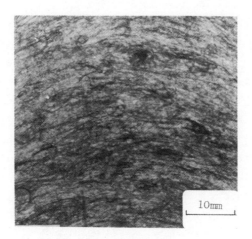

FIG. 5.5. Macroradiograph of a region in a glass fibre reinforced
polypropylene injection moulded disc.
(After Darlington et al, 1976).

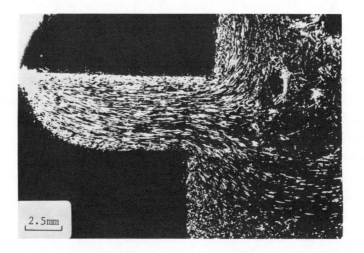

FIG. 5.6. Contact microradiograph of a thin section (50-100μm) taken
from a glass fibre reinforced polypropylene injection moulded
component.

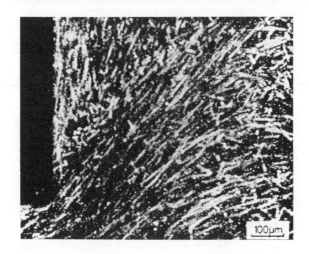

FIG. 5.7. Part of a microradiograph of a thin section cut from a
commercial injection moulding produced in pigmented, short glass
fibre reinforced thermoplastic polyester.
(After Darlington and McGinley, 1975, J.Mat.Sci., 10, 906-909)

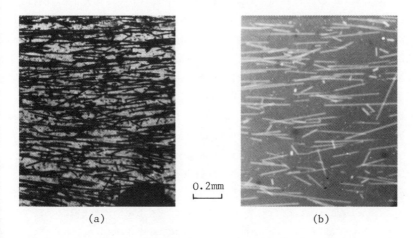

(a)   (b)

FIG. 5.8. Comparison of the micrographs obtained using transmitted
light (a) and X-rays (b) from a thin section cut from a short
carbon/glass reinforced nylon 66 hybrid composite.
(After Folkes and Sharp, 1981)

non-destructive and can also be applied to a study of the anisotropy arising in complex shaped components – see section 7.2.3. Measurement of the velocity of an ultrasonic wave in a material can give a measure of the elastic modulus in the direction of propagation – Musgrave (1954) and Markham (1970). In an isotropic material, the tensile modulus is equal to ρV_L^2 where ρ is the density and V_L the velocity of longitudinal waves. In an anisotropic material, the anisotropy can be assessed by measuring V_L^2 in various directions in the sample. A set of results for a dumbbell test specimen moulded in polytetramethylene-terephthalate and containing 30% by weight of glass fibres is shown in Fig. 5.9. The values of V_L^2 obtained through the thickness and width of the gaugelength are virtually identical, but are very different from the value obtained along the axis of the test bar. This shows that the high modulus glass fibres tend to lie along the axis of the bar, a result confirmed by contact micro-radiography – Fig 5.10.

5.1.5 Scanning electron microscopy

The techniques described so far are concerned primarily with the characterisation of fibre orientation. A vital consideration, though, in any composite material is the effectiveness of the bond existing between the fibres and matrix. One established method of approach is to fracture the specimen and then to study the fracture surface using the large depth of focus of the scanning electron microscope (SEM). For example, when the fibre-matrix bond is very poor, the fracture surface will consist of a large number of long fibres protruding from the surface, together with holes in the matrix from which fibres have been pulled during composite fracture – see Fig 5.11(b). When the bond is good and the composite contains long fibres, the fibres will break when the composite is fractured, so that the fracture surface will be relatively smooth. The longest fibre that is pulled out of a fracture surface is $\ell_c/2$, where ℓ_c is the critical fibre length $= \frac{\sigma_{uf}d}{2\tau_u}$. If ℓ_c is obtained from such measurements, then the interfacial bond strength τ_u may be derived. Apart from specific measurements of ℓ_c, qualitative conclusions regarding the fibre-matrix bond may be drawn from the degree of adhesion of the matrix to the

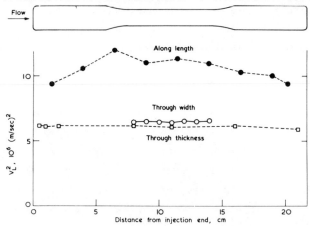

FIG. 5.9. Anisotropy of elastic modulus in an ASTM testbar moulded in glass fibre reinforced polytetramethyleneterephthalate. V_L^2 is the square of the 5 MHz ultrasonic wave velocity, which is a measure of the longitudinal modulus.

(After Thomas and Meyer, 1976)

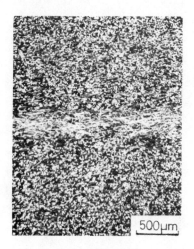

FIG. 5.10. Variation of fibre orientation through the thickness direction in a section cut transversely to the axis of an ASTM bar; glass fibre reinforced nylon 66.

(After Darlington and McGinley, 1975, J.Mat.Sci., <u>10</u>, 906-909)

fibres. An example of good bonding is shown in Fig. 5.11(a) and contrasts with the case of poor bonding shown in Fig. 5.11(b). SEM is now sufficiently routine that it can be used to assist in the optimization of the fibre surface treatment and sizing procedures that are fundamental to the successful exploitation of short fibres in thermoplastics.

In an oriented specimen of a short fibre reinforced thermoplastic the nature of the fracture surface will of course depend on the direction of the load with respect to the overall fibre direction. Thomas and Meyer (1976) have obtained SEM pictures of fracture surfaces of miniature Izod samples, cut longitudinally and transversely from an ASTM tensile bar, moulded in glass fibre reinforced poly-tetramethyleneterephthalate. Typical results that they obtained are shown in Fig. 5.12. The significant preferred orientation of the fibres along the original axis of the ASTM bar is notable. At lower magnification, it becomes clear that this preferred orientation is inhomogeneous - Fig. 5.13. The existence of a thin central region having transversely oriented fibres is apparent. This layer-type structure is much in evidence in moulded components and is indeed a basic consequence of the injection moulding process. We will return to this later in the chapter.

5.2 EFFECT OF MOULDING CONDITIONS ON FIBRE LENGTH AND ORIENTATION

5.2.1 Fibre length

During the injection moulding process, considerable fibre breakage can take place. This may easily result in a large population of fibres in the moulded component, having a length that is much too small to be really effective in ensuring good stiffness and strength of the composite. Since ℓ_c for many fibre-matrix combinations is ~200μ, it follows that fibres having a length in the range 1.5-2.0mm are ideally required in the final moulding. This Utopian situation is rarely achieved using injection moulding and may well be impossible, unless some radically different fabrication route is used. However, some modest degree of control of the fibre length is possible by manipulation of the moulding conditions. Filbert (1969) isolated some of the critical parameters in the moulding of glass fibre

(a)

(b)

FIG. 5.11. Scanning electron micrographs of fracture surfaces of
carbon fibre reinforced nylon 66 for the case of (a) good fibre-
matrix bonding (b) poor fibre-matrix bonding.

(a)

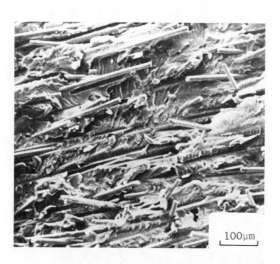

(b)

FIG. 5.12. Scanning electron micrographs of fracture surfaces of
miniature Izod impact specimens cut from an ASTM testbar, moulded
in glass fibre reinforced polytetramethyleneterephthalate, (a)
specimen cut longitudinally from the testbar (b) specimen cut
transversely from the test bar.
(After Thomas and Meyer, 1976).

FIG. 5.13. Scanning electron micrograph of the fracture surface of
a miniature Izod impact specimen cut trasversely to the axis of
an ASTM testbar, moulded in glass fibre reinforced polytetramethyleneterephthalate. The magnification in this micrograph
is lower compared to the previous figure and shows the presence of
an inhomogeneous fibre orientation across the sample.
(After Thomas and Meyer, 1976)

reinforced nylon 66:-

(a) <u>Rear zone temperature</u>. In common with results obtained during
the extrusion compounding of short fibres into thermoplastics, the
temperature at the feed zone of the screw has a relatively large
effect on fibre length. Table 8, taken from the work of Filbert
(1969), shows that as the rear zone temperature is increased, there
is an associated increase in fibre length. As pointed out by Kaliske
and Seifert (1975),the greatest increase in fibre length occurs when
this temperature approaches the melting point of nylon 66. They
suggest, therefore, that the outer layer of the granules will soften
and reduce the tendency for fibres to break, as a result of inter-
granular interaction. They recommend that the feedstock itself
be pre-heated to as high a temperature as possible and yet insufficient
to cause oxidative degradation, which for nylon 66 is about 80°C.

BARREL TEMP °C			SCREW RETRACTION TIME secs	WEIGHT AVERAGE FIBRE LENGTH μm	TENSILE STRENGTH MNm^{-2}	FLEXURAL MODULUS GNm^{-2}
REAR	CENTRE	FRONT				
288	274	274	16.5	556	184.80	9.76
274	274	274	16.8	544	183.55	9.67
260	274	274	18.9	513	182.03	9.49
246	274	274	20.5	500	181.00	9.44
232	274	274	21.1	495	179.75	9.35

TABLE 8. Effect of rear zone temperature, used during injection moulding, on the average fibre length and physical properties of ASTM tensile bars of short glass fibre reinforced nylon 66. (Data extracted from Filbert, 1969)

SCREW SPEED r.p.m.	BACK PRESSURE MNm^{-2}			
	0	8.96	17.93	26.89
48	556	541	SCREW WOULD NOT RETRACT	
96	493	483	472	432
186	452	445	427	386

TABLE 9. Weight average fibre length (μm) in ASTM tensile bars of glass fibre reinforced nylon 66 for various screw speeds and back pressures during injection moulding.
(Data extracted from Filbert, 1969)

(b) <u>Screw speed and back pressure</u>

One major source of fibre breakage occurs during screw - back. This
part of the cycle is concerned with the production of a fully
homogenized melt and so normally, for an unfilled thermoplastic,
this is aided by employing a high screw speed and back pressure.
These are not appropriate conditions to use when fibres are present
since excessive breakage will occur. The choice of conditions,
of course, are dictated by similar considerations to those in
extrusion compounding, namely the need to disperse the fibres without
breaking them. If the compounding stage has been accomplished
professionally and the feedstock for the moulding operation contains
well dispersed long fibres then, as Filbert (1969) has shown, fibre
length increases as screw speed and back pressure decrease. Table 9,
taken from this work, shows that in fact these two parameters have
a more dramatic effect on fibre length than changes to the rear zone
temperature. Increasing screw speed at any back pressure tends to
cause more fibre damage than increasing back pressure at any screw
speed. Filbert suggests that fibre breakage is more closely related
to the peripheral speed of the screw i.e. the wiping action generated
by the screw in the barrel undoubtedly causes fibres to break.
Flemming (1973) offers the alternative explanation that as the screw
speed increases, the melting zone is shifted a long way to the front
of the screw, where the shear rate is high. Substantial fibre length
reduction can then occur by the same kind of mechanism that arises
from having a low rear zone temperature. In view of these effects,
it is normal practice to employ zero back pressure and the slowest
screw speed that will produce the required shot in the allotted
overall cycle time.

In addition to these main process parameters, there are of course,
other factors which can exert an influence on fibre length in the
moulded components. Kaliske and Seifert (1975) give a long list of
all the factors that can contribute to fibre attrition. One in
particular is worthy of special note. From the work of Richards and
Sims (1971) it appears that, in the case of glass fibres, the type
of coupling agent used exerts a major influence on the properties of

the moulding. Furthermore, the variation of component properties
with, e.g. screw speed also depends on the particular surface
treatment – see Fig. 5.14. Although some of the observed effects
may simply be due to changes in the fibre-matrix bond strength,
it has been suggested that fibre dispersion and/or breakage may also
be varying with fibre surface treatment. The necessity of having
glass fibre surfaces treated in order to minimize damage has been
known for a long time. Other high modulus fibres, such as carbon,
used as reinforcement for thermoplastics, are also treated but
usually with the aim of improving the fibre-matrix bond. This raises
the interesting question as to whether more effort should be devoted
to the development of improved methods for protecting fibre surfaces
from damage during processing. Only a limited amount of control
over fibre breakage is possible by manipulation of the processing
conditions used during moulding, and so improvements in fibre
protection could prove complementary to this.

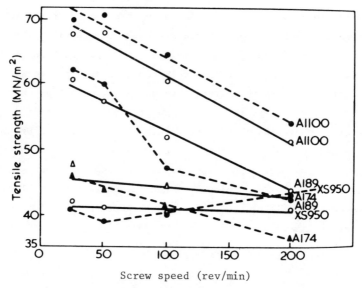

FIG. 5.14. Tensile strength versus screw speed for short glass fibre
reinforced polypropylene test bars. The data refers to the
injection moulding of a dry mix of glass fibres and polypropylene
powder. The continuous lines are for dry samples, broken lines
for samples which have been immersed in water at 70°C for one
day. The labels on the graphs refer to different silane
coupling agents.
(After Richards and Sims, 1971).

5.2.2. Fibre orientation

During the moulding process, the fibres may become oriented in a
complex manner. In the component itself, a characteristic "layer"
structure is frequently observed, with the fibres oriented in quite
different ways according to their location through the thickness -
Kaliske and Meyer (1975), McNally (1977). For short glass fibre-
filled thermosetting materials, it has been shown that flow geometry
is the major parameter affecting fibre orientation - Goettler (1970),
Owen and Whybrew (1976). The resin viscosity and flow rate can
significantly alter the proportions of the oriented regions.
Orientation that is unfavourable with respect to the loads encountered
in service can lead to component failure, both in thermosetting
and thermoplastic materials - Rowbotham (1974). Apart from the
flow geometry, there are a number of processing variables that can
affect the fibre orientation in any given reinforced thermoplastic.
The major variables are:-

(i)   Melt temperature

(ii)   Mould temperature

(iii) Injection speed

The moulding of thermoplastics involves non-isothermal flow i.e. the
injection of a melt into a relatively cold cavity. A skin layer
of solidified material will form at the walls of the cavity. The
thickness of this layer will be affected by all three of the above
variables e.g. a reduction in mould temperature will cause an
increase in skin thickness. If the fibres in the skin region are
oriented in a different manner to those elsewhere, then obviously
the anisotropy in the component will be altered by changes in the
above variables. We can see this clearly in the case of a simple
strip mould studied by Bright et al (1978). The mould used was
a two cavity bar mould (Fig. 5.15) of dimensions 190 x 30 x 1.5 mm
with a semicircular gate of 1mm radius. Mouldings were produced
in a polypropylene containing 20% by weight of short glass fibres.
Three sets of processing conditions were used:

(i)   Fast constant speed injection (injection time = 0.2 sec)

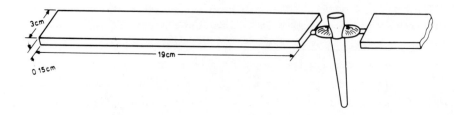

FIG. 5.15. Simple strip moulding used for fibre orientation studies
(After Bright et al, 1978)

(ii)   Slow constant speed injection (injection time = 11 sec)

(iii) Stepped injection, starting slowly and abruptly changing
to fast after 50% of the shot had been injected.

All other processing conditions were kept constant. Contact microradiographs showing fibre orientation within the moulding are given
in Fig 5.16. for fast injection and Fig.5.17. for slow injection.

In both cases a layered structure is observed. For fast injection,
the core of the moulding contains fibres mainly aligned perpendicular
to the flow direction. Above and below this are regions with the
predominant fibre orientation in the flow direction. Most of the
fibres are lying in the plane of the moulding. The orientation
pattern is markedly different for slow injection. Here the central
region is of greater thickness and contains fibres highly aligned
in the flow direction. Between the "skin" and "core" regions there
appears to be a fibre free layer. Note that if the change in fibre
orientation through the moulding thickness is used to define the
proportions of skin and core material, it would appear that the skin
thickness for fast injection is considerably greater than for slow
injection. This anomaly has been resolved by Folkes and Russell (1980)

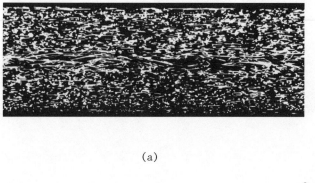

(a)

0.5mm

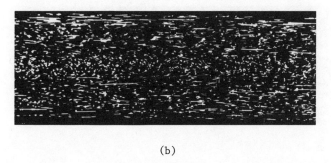

(b)

FIG. 5.16. Contact microradiographs of sections cut from the simple
   strip moulding shown in Fig. 5.15, (a) section cut perpendicular
   to the major flow direction for fast injection (b) section cut in
   the plane containing the major flow direction and the thickness
   direction for fast injection.
   (After Bright et al, 1978)

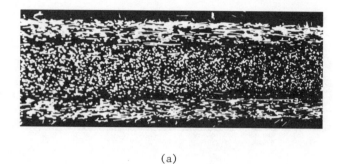

(a)

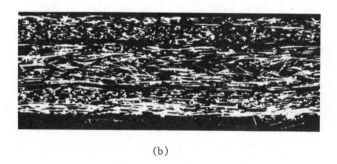

(b)

FIG. 5.17. Contact microradiographs of sections cut from the simple
strip moulding shown in Fig. 5.15, (a) section cut perpendicular
to the major flow direction for slow injection (b) section cut
in the plane containing the major flow direction and the thickness
direction for slow injection.
(After Bright et al, 1978)

using birefringence data. In the case of mouldings produced using
stepped injection, it was found that at the end of the moulding
nearest the gate, the pattern of orientation was similar to that
obtained for a moulding injected slowly throughout. At the far end,
the pattern was similar to that obtained by fast injection. The
change in orientation pattern along the bar was very abrupt.
Although the slow injection experiments were conducted using much
lower speeds than would be normal for commercial moulders, the
levels of shear rate obtained will occur, nevertheless, when moulding
articles containing thick cross-sections.

The orientation effects described above were in relation to a
parallel sided strip moulding. However, dumbbell shaped tensile
test specimens, moulded to ASTM standards, show a very similar
pattern of orientation to that produced in a parallel sided strip
using normal i.e. fast injection conditions. It is, of course, the
presence of a predominant fibre orientation along the axis of the
test piece , that has made it a popular specimen for obtaining an
upper bound for the tensile modulus of the composite. It is
important, though, to keep these ideas in perspective, since although
widely accepted as a standard test specimen, the fibres are not
fully aligned throughout the moulding cross-section — see Fig. 5.10.
Kaliske and Meyer (1975) have shown that the pattern of fibre
orientation observed for fast injection also extends to the case
where the specimen is an edged gated plaque. Tensile specimens cut
at 90° to the overall flow direction have been widely used as a
method of obtaining a measure of the lower bound of the tensile
modulus of the composite. However, this can be a dangerous procedure,
since changes in moulding conditions and materials can significantly
affect the anisotropy in such a plaque. This aspect will be
discussed more fully in Chapter 7.

Of course, practical moulded components will be much more complex
than those described above. It is of interest to see whether fibre
orientation distributions measured in idealised geometries can assist
in our appreciation of those measured in selected positions of
complex mouldings. It was to this end that Bright and Darlington

(1980) made a detailed study of a "display stand", moulded in glass fibre reinforced polypropylene. A general view of the moulding is shown in Fig 5.18., while the positions from which specimens were cut for fibre orientation studies are indicated in Fig 5.19. A diagramatic representation of the fibre orientation patterns corresponding to positions A, B, C and D are shown in Fig. 5.20. The main results may be summarised as follows:--

Position A This is a thin-walled section near to the sprue. The fibre orientation pattern is similar to that described above for a simple bar moulding for fast injection. Indeed, it appears to occur generally in those areas of the moulding, where the flow is not disturbed by other effects of geometry such as corners, junctions etc.

Position B This is a junction where the flow divides; one arm of the junction being approximately twice the thickness of the others. One particularly significant feature is the development of the very thick core of transversely aligned fibres in the thicker section. This also serves to show the importance of section thickness as a variable controlling anisotropy in moulded components.

Position C The flow path in this case passes around a right-angle corner with a simultaneous reduction in section thickness of about 50%. The layers of fibres aligned near the surface appear to find their own radius around the bend, with the outer region of the corner containing fibres aligned predominantly transverse to flow. This is a common feature whenever short fibre reinforced thermo- plastics flow close to sharp corners. Surprisingly, as the thickness reduces around the corner, there is no marked increase in the degree of fibre alignment parallel to flow, as is often observed in converging flow. The main change is simply a reduction in core thickness.

Position D This is a section taken through a shallow reinforcing rib. The flow is distorted by the rib, but again the layer of fibres aligned in the flow direction does not follow the comparatively sharp rib contour. At the tip of the rib, the fibres are relatively

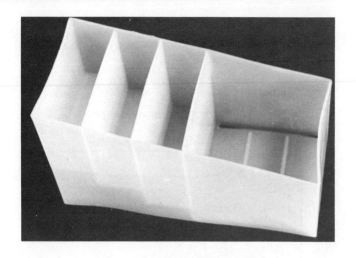

FIG. 5.18.   General view of display stand moulding.

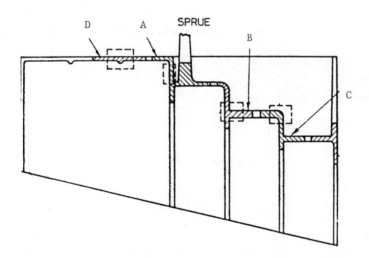

FIG. 5.19.   Display stand moulding, showing the positions from which
thin sections were cut for a study of fibre orientation.
(After Bright and Darlington, 1980)

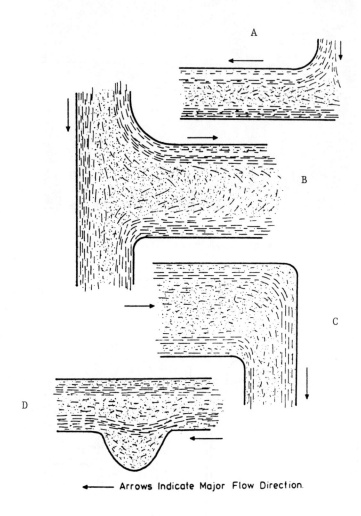

A

B

C

D

◄──── Arrows Indicate Major Flow Direction.

FIG. 5.20. Diagrammatic representation of the fibre orientation
distributions at the various locations in the display stand
moulding defined in the previous diagram.
(After Bright and Darlington, 1980)

well aligned transverse to the overall flow direction.

Bright and Darlington (1980) conclude that the effects of mould geometry on fibre orientation distribution tend to be much greater than the effects of moulding conditions. It is often the case that the fracture of a plastics component will be initiated where flow singularities occur. Since the type of flow that arises is dependent on the geometry of the particular component, gating arrangements etc, it is important to realise that simple moulded test specimens are no real substitute for a thorough examination of the actual component itself.

When more than one gate is used, additional problems may arise due to the formation of weld lines. Basically these can be of two types:-

    (i) Butt welds - where two separate flow fronts meet head-on. A simple example is a straight bar which is moulded using a gate at both ends, usually referred to as double end gating.

    (ii) Knit line - where one advancing flow front is split by some obstacle in the path of the flow and then the separate flow fronts recombine. A good example of this arises when metal components, such as bosses, are moulded into the plastics component.

In both cases, the weld line will be a potential source of weakness in the component. Material on the surface of the flow fronts will tend to degrade during mould filling, so that where they meet there will be incomplete mixing of the melt. Since the fibres tend to lie parallel to the surface of the advancing flow fronts, they will be oriented parallel to the weld line, as shown in Fig. 5.21. The fractional loss in strength at the weld line is greatly increased by the presence of fibres and of course results primarily from the improper orientation of the fibres. Cloud and McDowell (1976) state the need for adequate mould venting and prolonged hold times, in order to maximize the weld line strength, but as usual this can only be done at the expense of increasing the overall cycle time.

5.3. SURFACE FINISH OF MOULDED COMPONENTS

The surface finish produced on a moulded component may appear to be of trivial interest, but it is certainly of considerable importance to commercial moulders. In unfilled thermoplastics, surface finish is rarely a real problem and glossy mouldings can usually be produced providing the mould cavity is well finished. In general, the quality of the surface finish of components moulded in short fibre reinforced thermoplastics is inferior to their unfilled counterparts. There are of course exceptions e.g. electric drill housings have an excellent surface finish. To a large extent, though, the need to achieve really good finishes in reinforced thermoplastics is less important, since many of these components are used for engineering as distinct from domestic applications. In the case of short glass reinforced thermoplastics, surface finish has been related to the injection speed and/or pressure used during the moulding operation - Bright et al (1978). This has been demonstrated rather clearly by producing mouldings using stepped injection e.g. starting with slow injection and abruptly changing to fast after 50% of the shot has been injected. Fig. 5.22 shows a scanning electron micrograph of the matt to gloss transition on the surface of such a moulding. It can be seen that there is a very sharp transition in surface finish. The end of the moulding nearest the gate has the matt surface i.e. the slow injection portion. This matt region contains many ridges, which appear to be individual fibres coated with polypropylene. This is shown at greater magnification in Fig. 5.23. The effects observed are believed to be related to the time taken for a significant pressure to develop in the mould cavity. It is assumed that the material deposited on the cavity wall during injection has initially an appearance similar to that shown in Fig. 5.23., with individual melt covered fibres protruding from the surface. For this to be transformed to a smooth gloss surface, sufficient pressure must be available in the mould cavity to push this material firmly into contact with the wall, and to force the polymer melt into the gaps between surface fibres. The most favourable condition for this to occur will be when the cavity pressure is high and the skin layer is thin i.e. fast injection.

There may be cases, of course, where it is desirable to have fibres protruding from the surface. As an example, carbon fibre reinforced thermoplastics can be used in dry bearing applications, but since it is the low coefficient of friction of the fibre that is being exploited and not a property of the matrix, it is essential that the fibres are at the surface of the moulding and not covered by a layer of matrix.

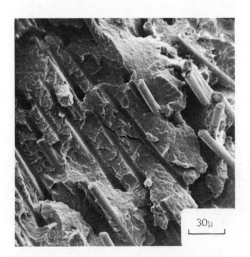

FIG. 5.21. Scanning electron micrograph of a fracture surface in carbon fibre reinforced nylon 66 taken in the vicinity of a weld line; ASTM test bar moulded using double-end gating.

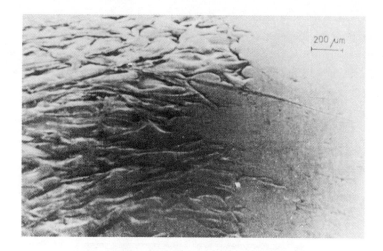

FIG. 5.22.   Scanning electron micrograph of the matt-to-gloss
transition on the surface of an injection moulded strip of glass
fibre reinforced polypropylene.
(After Bright et al, 1978)

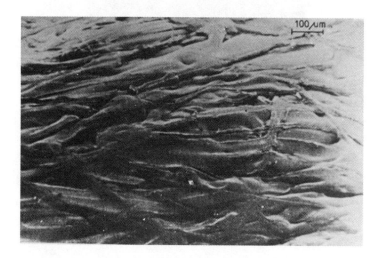

FIG. 5.23.   Scanning electron micrograph of the matt surface region
shown at higher magnification compared to the previous figure.
(After Bright et al, 1978)

References

Allan, P. and Bevis, M.J. (1981). Unpublished work.

Bright, P.F. (1980). Factors Influencing Fibre Orientation and Product Performance in Fibre Reinforced Thermoplastics Injection Mouldings. Ph.D. Thesis, Department of Materials, Cranfield Institute of Technology, Bedford.

Bright, P.F., Crowson, R.J. and Folkes M.J. (1978). A study of the effect of injection speed on fibre orientation in simple mouldings of short glass fibre-filled polypropylene. J.Mat.Sci., 13, 2497-2506.

Bright, P.F. and Darlington, M.W. (Nov 1980). The structure and properties of glass fibre filled polypropylene injection mouldings. Plastics and Rubber Institute Conference "Moulding of Polyolefins", London.

Cloud, P.J. and McDowell, F. (Aug 1976). Reinforced thermoplastics: Understanding weld-line integrity. Plastics Technology.

Darlington, M.W., McGinley, P.L. and Smith, G.R. (1976). Structure and anisotropy of stiffness in glass fibre reinforced thermoplastics. J.Mat.Sci., 11, 877-886.

Darlington, M.W. and Smith, G.R. (1975). Voiding in glass fibre reinforced thermoplastics mouldings. Polymer, 16, 459-462.

Engstrom, A. (1956). Physical Techniques in Biological Research. ed. G. Oster and A.W. Pollister, vol 3, p495. Academic Press, New York.

Filbert, W.C. (1969). Reinforced 66 nylon - molding variables vs fibre length vs physical properties. SPE Journal, 25, 65-69.

Flemming, F. (1973). The influence of processing techniques on the properties of fibre-reinforced thermoplastics. Plaste u Kaut., 20, 767-772.

Folkes, M.J. and Russell, D.A.M. (1980). Orientation effects during the flow of short fibre reinforced thermoplastics. Polymer, 21, 1252-1258.

Folkes, M.J. and Sharp, J. (1981). To be published.

Goettler, L.A. (April 1970) Controlling flow orientation in molding of short-fiber compounds. Modern Plastics, 140-146.

Hunter, W.A. (Feb. 1975). The fine points of processing glass-reinforced engineering resins Plast. Eng., 31, 24-28.

Kaliske, G. and Meyer, F. (1975). The influence of fibre orientation in injection moulded glass fibre reinforced thermoplastics of the short fibre type on the anisotropy of mechanical properties. Plaste u Kaut, 22, 496-498.

Kaliske, G. and Seifert, H. (1975). The extent to which fibre breakdown can be influenced when injection moulding glass fibre reinforced thermoplastics - illustrated with reference to the processing of Miramid VE30. Plaste u Kaut., 9, 739-741.

Lucius, O.W. (1973). Practical experience in the injection moulding of glass fibre reinforced thermoplastics. Kunstoffe, 63, 367-372.

Markham, M.F. (1970). Measurement of the elastic constants of fibre composites by ultrasonics. Composites, 1, 145-149.

Maxwell, J. (1964). The processing of glass-filled nylon. Plastics Today, 22, 9-13.

McNally, D. (1977). Short fiber orientation and its effect on the properties of thermoplastic composite materials. Polym-Plast. Technol. Eng., 8(2), 101-154.

Murphy, T.P. (June 1965). How to mold FRTP resins. Mod. Plast., 42, 127.

Musgrave, M.J.P. (1954). On the propagation of elastic waves in aeolotropic media: I General principles. Proc.Roy.Soc., A226, 339-355.

Nemet, A., Black, A.D. and Cox, W.F. (1962). Xeroradiography for the testing of plastics and light materials. Trans. J. Plast. Inst., 30, 192-198.

Owen, M.J. and Whybrew, K. (Dec. 1976). Fibre orientation and mechanical properties in polyester dough moulding compounds (DMC). Plastics and Rubber, 1, 231-238.

Pye, R. (1968). Injection Mould Design for Thermoplastics. Iliffe
    Books Ltd, London.

Richards, R.W. and Sims, D. (Dec 1971). Glass-filled thermoplastics –
    effects of glass variables and processing on properties.Composites,
    214-220.

Rowbotham, E.M. (1974). Fiber orientation in fiber-reinforced
    plastics and how it affects automotive applications. S.A.E.
    Automotive Engineering Congress, Detroit, Michigan, reprint 740264.

Schlich, W.R., Hagan, R.S., Thomas, J.R., Thomas, D.P. and Musselman,
    K.A. (Feb. 1968). Critical parameters for direct injection molding
    of glass-fiber-thermoplastic powder blends. S.P.E. Journal, 24,
    43-54.

Theberge, J.E. (1973). How to process glass fiber fortified
    thermoplastics. Plast. Des. Process, Part I, January, pp14-20.
    Part II, February, pp18-20.

Thomas, K. and Meyer, D.E. (Sept 1976). Study of glass-fibre-
    reinforced thermoplastic mouldings. Plastics and Rubber: Processing,
    99-108.

CHAPTER 6
Rheology

In the previous chapter, a description was given of some of the work
that has been carried out aimed at relating the processing conditions
used during moulding to the microstructure of the component. It is
clear that the final orientation adopted by the fibres, at the
conclusion of the moulding cycle, will have a major effect on the
anisotropy of the physical properties of the moulded component. The
degree of anisotropy may be considerable, compared to an equivalent
moulding in an unfilled thermoplastic. Hence, it is desirable to be
able to make even qualitative predictions of the fibre orientation
from one part of a moulding to another. Since tooling costs for
comparatively simple mouldings can be very large, it is vital that
the mould designer can anticipate the likely effects of gating, changes
of cross-section etc. on the fibre orientation distribution. In this
way, the optimum fibre orientation distribution can be developed for
any particular application. The day when quantitative predictions of
fibre orientation in complicated shaped mouldings can be made is still
a long way off and, as Lockett (1980) has indicated, the effort
involved may not be justified. In most cases what is needed are
general guidelines concerning the behaviour of fibres while moving in
various geometries, from which their behaviour in a more complex
geometry can be synthesised.

This chapter is concerned therefore with two distinct but related
problems; one being the characterisation of the basic rheological
properties of short fibre reinforced thermoplastics and the second,

an examination of the rheology of mould filling during injection moulding.

6.1 BASIC RHEOLOGICAL BEHAVIOUR

The rheological properties of short fibre reinforced themoplastics may differ in detail from those of normal unfilled polymers but they are not grossly different. This is perhaps rather surprising in view of the fact that we are concerned here with viscoelastic fluids often containing very high concentrations of fibres. Furthermore, it should be remembered that formally we are examining a "structured" melt and one for which the rheological parameters e.g. viscosity are expected to be anisotropic. This latter point is clearly important to theoretical rheologists, who are concerned with the formal representation of rheological parameters for anisotropic fluids – the counterpart in fact of anisotropic elasticity theory. In cases, where one is attempting to use rheological data to explain major fibre orientation effects in moulded components, the rigorous formalism is usually disregarded. One then assumes that the normal approach to the measurement of rheological parameters can be adopted, with only minor modification to the experimental procedures.

In general, two types of flow can be distinguished, namely shear and extensional (irrotational) flow and for each, there will be a characteristic viscosity and elasticity parameter. In principle then, there will be four viscometric parameters to be measured. In practice, it is the shear viscosity that receives the most attention,while elasticity is usually inferred from die-swell data. Shear viscosity can be measured using most of the normal techniques developed for unfilled melts i.e. cone and plate, capillary, dynamic mechanical etc. In the former case, some problems of sample rejection from the gap can arise, but this can be overcome using a biconical rotor in an enclosed cavity, in which the melt can be subjected to a modest hydrostatic pressure. In view of the similarity of the type of flow involved with mould filling, it is capillary rheometry that is most frequently employed for characterising the shear viscosity of short fibre filled thermoplastics.

6.1.1. Measurement of Apparent Viscosity

Most of the data for short fibre filled thermoplastics, reported in
the literature, have been obtained using a constant volume flow rate
capillary rheometer, of the kind manufactured by Daventest Ltd and
Instron Ltd. This is a convenient and comparatively simple method,
in principle, for obtaining viscosity data, and consists of measuring
the pressure drop along a die for different values of the volume
flow rate. A detailed description of the techniques of capillary
rheometry is given by Van Wazer et al (1963).

The usual expressions used to relate pressure drop ΔP to shear stress
τ and volume flow rate Q to Newtonian shear rate $\dot{\gamma}_N$, in a capillary
of radius R and length L are as follows:

$$\tau = \frac{\Delta PR}{2L} \quad \ldots\ldots(6.1), \quad \dot{\gamma}_N = \frac{4Q}{\pi R^3} \quad \ldots\ldots(6.2)$$

where both τ and $\dot{\gamma}_N$ apply at the capillary wall. In practice, both
equations require correction. Equation (6.2) may be corrected using
the Rabinowitsch correction, which makes allowance for the fact
that the material may be non-Newtonian. This correction is given by:-

$$\dot{\gamma} = \frac{1}{4} \left\{ 3 + \frac{d \log \dot{\gamma}_N}{d \log \tau} \right\} \dot{\gamma}_N \qquad \ldots\ldots\ldots(6.3)$$

where $\dot{\gamma}$ is the true shear rate. The expression for shear stress is
also incorrect, because it assumes that the pressure drop along the
die is linear. This is not usually the case and quite significant
contributions to the total pressure drop along the die arise because
of entrance and exit effects. This correction to equation (6.1) may
be made in several ways. One approach is merely to use a die with
large L/D, where D is the diameter, and assume that the additional
pressure required in the die entrance region will then be only a
small proportion of the total pressure drop (Method 1). An alternative
technique (Method 2), used by many workers, is to make measurements
on two dies of different length, but equal diameter, and subtract
the pressure drops corresponding to a given flow rate. Often the
shorter die has a length that is much smaller than the die diameter,

in which case it is referred to as a "zero length" die. Equation
(6.1) then becomes:-

$$\tau = \frac{\left(\Delta P_1 - \Delta P_2\right) R}{2 (L_1 - L_2)}$$

where ΔP_1 and ΔP_2 denote pressure drops along dies of length L_1 and
L_2.

A third method of correction (Method 3) is the one usually ascribed
to Bagley (1957). In this method, a fictitious die of length L + ND
(N is dimensionless) is considered, along which there is a linear
pressure drop. One may then write:-

$$\tau = \frac{\Delta P \ D}{4 (L + ND)}$$

and hence $\Delta P = 4\tau \left\{ \dfrac{L}{D} + N \right\}$

Measurements of the pressure drop ΔP for a number of dies with
different L/D, for fixed shear rates, will give a straight line,
from which τ may be evaluated.

For an accurate assessment of the apparent viscosity $\eta (= \frac{\tau}{\dot\gamma})$ and its
dependence on shear rate or shear stress, together with the possible
effects of material parameters etc., it is important to take heed
of the above correction procedures, even though they are tedious
to apply.

6.1.2. Experimental Observations

Work on the rheology of short fibre reinforced thermoplastics dates
from the mid 1960's, but the broader subject of the rheology of
dispersions developed from the work of Einstein (1906), who considered
the case of dilute suspensions of spherical fillers in Newtonian
fluids. Jeffery (1922) derived an expression for the shear viscosity
of a suspension of ellipsoidal - shaped particles and recently the
extensional flow of a suspension of fibres has also been considered -
Mewis and Metzner (1974). However, much of the earlier work on the
rheology of dispersions was concerned with very dilute systems, for

which particle-particle interactions could be neglected. The
rheology of these types of dispersions will differ significantly
from that of filled thermoplastics of commercial importance, where
individual particles are in close proximity to each other and the
host fluid is highly visco-elastic. The rather special rheological
properties of these materials manifest themselves in a number of
ways. Fig. 6.1 shows a log-log plot of shear viscosity versus shear
rate for three grades of polypropylene, containing 0, 20 and 30% by
weight of short glass fibres respectively, and having a modal length
of about 350-500μm. The unfilled material is almost Newtonian at
low shear rates, but as the shear rate increases the material becomes
increasingly pseudoplastic. At low shear rates,the presence of the
fibres causes an appreciable increase in viscosity, but it is note-
worthy that the viscosity values for the three materials converge to
a very similar value, at a shear rate in the range 10^4-10^5 sec^{-1}.
Essentially the same pattern of behaviour is shown by nylon 66 - see
Fig. 6.2. Similar shaped flow curves have been observed for other
fibre filled polymers - Wu (1979), Chan et al (1978) and Thomas and
Hagan (1966), but Charrier and Rieger (1974) saw little effect on
the flow curves due to the presence of the fibres. The similarity
in viscosity values at high shear rates, for filled and unfilled
thermoplastics, is an important factor in explaining the successful
exploitation of these materials in injection moulding technology,
since very little additional power will be required to mould the
filled materials.

The slopes of the flow curves shown in Fig. 6.1 converge to a value
close to -1 at high shear rates. It can be shown quite easily
(see e.g. Bernhardt (1959)) that this corresponds to a very blunt
velocity profile i.e. at these high shear rates, most of the shearing
flow takes place at the die wall, while the central body of material
is relatively unsheared. Some workers e.g. Goldsmith and Mason
(1967) have also observed radial migration of filler particles
towards the capillary axis during shear flow. If this occurs during
the flow of fibre filled thermoplastics, the region where most of
the shear takes place may be virtually fibre free. This could be a
contributory factor in explaining the very small dependence of

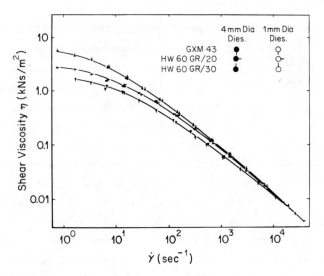

FIG.6.1. Apparent viscosity versus shear rate for three polypropy-
lenes containing different amounts of short glass fibres:-
GXM43 - polypropylene containing 0% weight fraction of fibres
HW60GR/20 - polypropylene containing 20% weight fraction of fibres
HW60GR/30 - polypropylene containing 30% weight fraction of fibres
(After Crowson and Folkes, 1980)

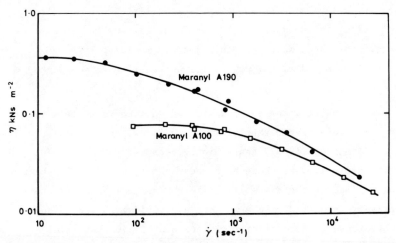

FIG. 6.2. Apparent viscosity versus shear rate for two nylon 66
materials containing different amounts of short glass fibres:-
Maranyl A100 - nylon 66 containing 0% weight fraction of fibres
Maranyl A190 - nylon 66 containing 33% weight fraction of fibres
(After Crowson and Folkes, 1980).

viscosity on fibre concentration at high shear rates. From computed
velocity profiles, derived from viscosity versus shear rate data,
the region over which fibre migration was taking place would have
to be several hundred microns thick if this argument is to be valid –
see Fig. 6.3. In this case , the fibre free layer would be clearly

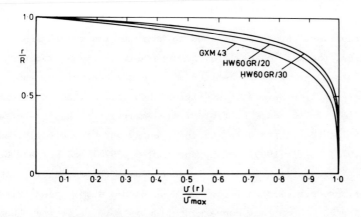

FIG. 6.3. Velocity profiles for three polypropylene materials at
a wall shear stress of 150 kNm^{-2}:–
GXM43 – polypropylene containing 0% weight fraction of fibres
HW60GR/20 – polypropylene containing 20% weight fraction of fibres
HW60GR/30 – polypropylene containing 30% weight fraction of fibres

(After Crowson and Folkes, 1980)

visible in a micrograph of the extrudate. This does not appear to be
the case for the glass fibre filled polypropylene and nylon 66
samples studied by Crowson and Folkes (1980), who proposed an
alternative explanation for the near coincidence of the viscosity
values at high shear rates, based a fibre-fibre interaction argument.
However, Wu (1979) has reported that migration phenomena occur
during the extrusion of glass fibre filled polyethylene terephthalate.
He shows that at low wall shear rates there is no migration, while
at intermediate shear rates, the fibres deplete at a relative radial
position of $r/R = 0.63$. This effect was originally observed by
Segre and Silberberg (1962) for neutrally buoyant spheres in
Newtonian fluids and referred to as the "tubular pinch effect". At

high wall shear rates, Wu (1979) observes a radial migration of
fibres towards the axis. However the resulting fibre concentration
distribution does not alter abruptly at the radial position
corresponding to a significant change in fluid velocity, and is
therefore unable to explain fully the viscosity versus shear rate
data at high shear rates.

When a capillary rheometer is used to characterize rheological
properties, the melt passes from a comparatively large diameter
barrel into a narrow cross-section die. Hence the melt is subjected
to convergent flow before it undergoes shear flow along the length
of the die. In the case of fibre filled melts, this has the effect of
pre-aligning the fibres prior to their entry into the die - Nicodemo
et al (1973), Bell (1969), Murty and Modlen (1977) and Lee and George
(1978). These observations are in accord with theoretical
expectations, e.g. Modlen (1969) has developed a simple geometrical
theory and has shown that convergent flow is more effective than
shear flow in aligning fibres. Subsequent shear flow along the die
can cause a significant misalignment of the fibres, and the fibre
orientation distribution at the exit of the die is a function of
the flow rate and the value of L/D. This is illustrated in Fig. 6.4.,
which shows longitudinal sections taken from glass fibre reinforced
polypropylene extrudates. The most pronounced fibre alignment occurs
when a very short die and high flow rate are used. This is consistent
with the observations made by Wu (1979), who predicts that higher
fibre alignment occurs with increasing pseudoplasticity of the melt.
These effects have been exploited by Parratt (1972) for the production
of aligned felts of short fibres, for use in thermoset mouldings - a
technique known as the PERME process.

Intuitively one would expect that the presence of highly aligned
fibres at the die entrance should have a significant effect on the
Bagley entrance correction factor, N, as compared to an unfilled melt.
Reported data however appears contradictory: According to Newman and
Trementozzi (1965) the end correction decreases with addition of
fibres while Roberts (1973) and Chan et al (1978) report to the
contrary. More recently, Crowson et al (1980) have shown that for

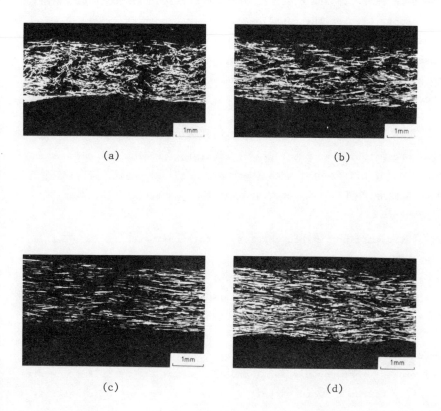

(a)

(b)

(c)

(d)

FIG. 6.4. Contact microradiographs of extrudates of glass fibre
reinforced polypropylene from a capillary rheometer. Extrudates
were obtained from dies of 2mm diameter, (a) 100mm long die, shear
rate = 1.5 sec^{-1}, (b) 100mm long die, shear rate = 24.0 sec^{-1},
(c) 0.3mm long die, shear rate = 24.0 sec^{-1}, (d) 0.3mm long die,
shear rate = 1.43 x 10^3 sec^{-1}.

(After Crowson et al, 1980)

small die diameters (< 1mm), non-linear Bagley plots can occur for
both filled and unfilled materials. These data for an unfilled
polypropylene and one containing 20% by weight of glass fibres are
shown in Figs 6.5 and 6.6. It seems unlikely that fibre orientation
is responsible for the non-linearity, since both materials show
similar behaviour. The data is consistent with a pressure dependent
viscosity i.e. as L/D is increased, larger pressures are required
to produce a given shear rate. The increase in pressure can
produce an increase in viscosity, so that the viscosity of the
material in the entrance region of a long die is greater than that
in a short die at the same shear rate. In most cases, it appears
that the established procedure of obtaining rheological data using
two dies of different length (Method 2) is applicable, providing
dies having an L/D < 20 are used and the diameter of the dies
exceeds 1mm.

As stated earlier in this chapter, a full characterization of the
flow properties of a viscoelastic fluid should include a knowledge
of its elastic behaviour. The normal indications of melt elasticity
are the existence of large die swells in capillary rheometry, a
large Weissenberg rod climing effect and large normal stress
differences, as measured for example using a Weissenberg rheogonio-
meter. Although there is continuing debate regarding the exact
relationship between die swell and the normal stress differences it
is nevertheless a rapid, albeit qualitative, measure of elasticity
which is widely used and is especially relevant in the case of fibre
filled thermoplastics. One reason for this is that the magnitude
of the die swell can affect the rheology of mould filling during
injection moulding - a matter which will be discussed in the next
section. In general, there seems to be almost universal agreement
that the incorporation of fibres causes a very significant reduction
in die swell of the parent melt - see e.g. Newman and Trementozzi
(1965), Roberts (1973), Chan et al (1978), Wu (1979). At low shear
rates, the true die swell is close to unity but since measurements
are usually made on the solidified extrudate, the actual die swell
appears to be less than unity because of thermal contraction. This
compares with values in excess of 3 and 4 for unfilled thermoplastics.

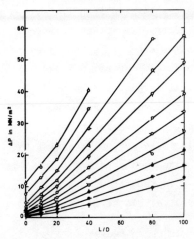

FIG. 6.5.   Pressure drop (ΔP) versus die aspect ratio (L/D) for 1mm
diameter dies.   Unfilled polypropylene at the flowing flow rates:
♦, 1130mm^3 sec^{-1};   ♂, 567mm^3 sec^{-1};   ☛, 283mm^3 sec^{-1};
☖, 142mm^3 sec^{-1};   ♀, 75.6mm^3 sec^{-1};   ♂, 37.8mm^3 sec^{-1};
☊, 18.9mm^3 sec^{-1};   ♭, 9.5mm^3 sec^{-1};   ♠, 4.7mm^3 sec^{-1};
☚, 2.4mm^3 sec^{-1};   ♠, 1.2mm^3 sec^{-1}.

(After Crowson et al, 1980)

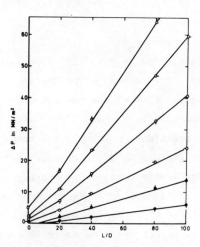

FIG. 6.6.   Pressure drop (ΔP) versus die aspect ratio (L/D) for 1mm
diameter dies; polypropylene containing 20% by weight of short
glass fibres.   For an explanation of the symbols see previous
figure.   (After Crowson et al, 1980).

However, depending on the particular polymer, the die swell of the
filled material can increase to values comparable to that of the
unfilled material, as the shear rate increases. Data for polypropylene and nylon 66 are shown in Figs. 6.7 and 6.8. As for apparent
viscosity, the die swell is also very dependent on the ratio of the
fibre length to the die diameter. For example, if the fibre length
exceeds roughly twice the die diameter and a "zero length" die is
used, the extrudate is so large and irregular that a meaningful
measure of die swell cannot be obtained. This phenomenon, though,
can be commercially exploited for use during the injection of cavity
wall insulation material - Cole et al (1979).

6.2 FLOW BEHAVIOUR IN COMPLEX GEOMETRIES

The foregoing discussion was concerned with some of the more
fundamental rheological properties of fibre-filled thermoplastics
and which have important implications for successful technological
processing. The flow geometry was deliberately chosen to be
straightforward, so that comparatively simple analytical relationships
can be used to extract rheological parameters from flow rate and
pressure drop data. In practice, however, the flow geometries
occurring in technological processing equipment will be very complex.
Furthermore, in the particular case of injection moulding, the fibre
filled melt will be subjected to very high flow rates under non-
isothermal conditions. The problem of predicting the fibre orientation
distribution throughout the final moulded component is formidable.
Lockett (1980) has made the very realistic comment that "it is
probably impossible, and almost certainly not economically viable, to
establish a procedure for making detailed predictions for moulds of
complex geometry under a wide range of processing conditions". What
can be done, though, is to develop a general idea of the fibre
orientation distribution, by combining the results obtained from a
number of simpler flow geometries. We may categorize these as:-

(i)   Shear flow - as will occur in a straight tube.

(ii) Convergent flow - (also referred to as extensional or
irrotational flow) - in the simplest case, this will occur when a

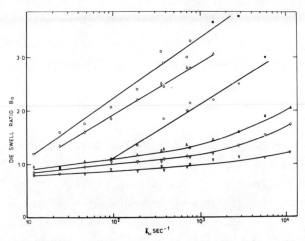

FIG. 6.7. Die swell ratio versus shear rate for two polypropylene
materials at 180, 210 and 270°C:

◻ Unfilled polypropylene at 180°C

♠ Unfilled polypropylene at 210°C

○ Unfilled polypropylene at 270°C

△ Polypropylene containing 20% wt fraction of glass fibres at 180°C

♀ Polypropylene containing 20% wt fraction of glass fibres at 210°C

▽ Polypropylene containing 20% wt fraction of glass fibres at 270°C

(After Crowson and Folkes, 1980).

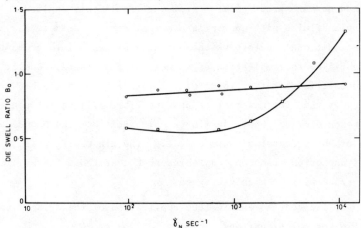

FIG. 6.8. Die swell ratio versus shear rate at 280°C, ◻ unfilled
nylon 66; ○ nylon containing 33% wt fraction of short glass fibres.
(After Crowson and Folkes, 1980).

a fluid passes from a wide to a narrow cross-section.

(iii) <u>Divergent flow</u> - in the simplest case, this will occur when a fluid passes from a narrow to a wide cross-section.

We have already seen that convergent flow leads to an alignment of fibres parallel to the flow direction. Divergent flow, on the other hand, imposes a uniaxial compression on the fluid elements parallel to the flow direction, since a deceleration must take place. This tends to orient the fibres orthogonally to the flow direction. Experimental confirmation of this pattern of behaviour arises from the work of Goettler (1970) and Owen and Whybrew (1976), using thermosets. Similar observations were reported for fibre filled thermoplastics in the previous Chapter. The situation with regard to shear flow is less clear cut. Goldsmith and Mason (1967) have shown that isolated fibres in Newtonian fluids undergoing simple shear flow will excecute a tumbling motion, with angular velocity equal to the shear rate. The establishment of a preferred fibre orientation along the flow direction, as frequently observed for fibre filled thermoplastics, depends on the existence of stable equilibrium positions for the fibres. Lockett (1980) has shown that a stable fibre orientation can arise as a result of a balance between viscous and normal stress effects i.e. the result only applies to non-Newtonian fluids with appropriate normal stress parameters. There are additional factors which must also be taken into account for fibre filled thermoplastics of commercial importance. One of the most important is that the fibres will interact with each other in such a way that the movement of any one fibre will be influenced by the presence of neighbouring fibres. In addition, the boundary surfaces of the processing equipment, e.g. the die walls, will impose an underlying orientation on the fibres such that those fibres close to the surface will be highly aligned.

It is now of interest to examine the application of these ideas to a number of practical situations.

6.2.1. <u>Jetting of melts during injection moulding</u>

The process of injection moulding involves the very rapid (often <1

sec) injection of a hot melt through a small gate into a comparatively
cold mould. The detailed process whereby the mould is filled is
dependent on the polymer being used and the processing conditions.
Some melts enter the cavity as a jet and this may frequently result
in the cavity being filled in the reverse direction to that
anticipated, as shown in Fig. 6.9. Unlike the process of uniform
mould filling, depicted in Fig. 6.10, jetting is very undesirable
since multiple weld-lines are generated in the completed moulding,
with a resulting deleterious effect on physical properties. In
those cases, where the precise location of the gate is not critical,
it is established practice to use side-gating so that the jetting
phenomenon only extends across the width of the moulding, ensuring
that the bulk of the mould is filled uniformly. However, in complex
mouldings, this practical method of minimising jetting may not be
possible, due to other considerations limiting the gate location.
It is desirable, therefore, to have available some criterion as to
whether jetting is likely to occur in any particular instance.
According to Oda et al (1976), the criterion for jetting is that the
melt emerging from the gate should not touch any surface of the
mould cavity, i.e. $B_oD \leqslant T$, where B_o is the die swell ratio for a
short die, D is a characteristic dimension of the gate e.g. depth or
diameter and T is the moulding thickness. This criterion has been
used to compare the mould filling behaviour of a number of polymers
with their corresponding die swell versus shear rate relationships
- Crowson and Folkes (1980). Figs. 6.7 and 6.8 show the die swell
of filled and unfilled polypropylene and nylon 66 versus shear rate
for a short ("zero-length") capillary. At the shear rates which
occur at the gate during injection moulding ($>> 10^4 \text{ sec}^{-1}$), the
two polypropylenes and the unfilled nylon 66 will have a die swell
sufficiently in excess of 1 to ensure that the melt contacts both
surfaces of the mould, for the gate sizes used in this particular
study. On the other hand, the fibre filled nylon, from the evidence
of the die swell studies will not. Consequently, this is the only
material that exhibited jetting during moulding.

6.2.2. Rheology of polymer melts during mould filling
On the assumption that jetting is normally avoided for the production

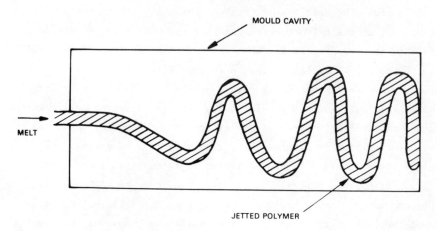

FIG. 6.9. Schematic diagram showing the jetting phenomenon during
injection moulding.

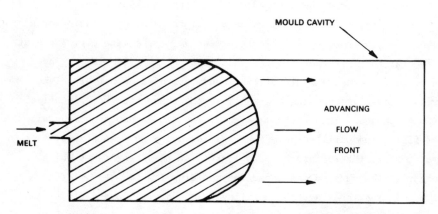

FIG. 6.10. Schematic diagram showing the process of uniform mould
filling during injection moulding.

of good quality mouldings, it is now necessary to examine the flow processes that take place and which are responsible for the final fibre orientation distribution in a moulding. We will assume that the mould fills uniformly, as depicted in Fig. 6.10 i.e. the melt progresses through the mould rather like an advancing wavefront. The discussion will be restricted to the events that occur in a simple rectangular mould, but the next section will consider some of the special problems associated with the even more complex situation existing when ribs, moulded in bosses etc feature in the moulding.

The moulding process involves the injection of a melt into a relatively cold cavity and so, during its passage through the cavity, solidification of the material must occur close to the boundary walls. One consequence of this is that the material which first enters the cavity does not gradually move forward to the far end of the cavity but rather forms solidified skins, through which further hot material must pass as the injection process proceeds. A schematic diagram of this process is shown in Fig. 6.11.

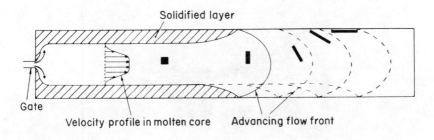

FIG. 6.11. Schematic diagram of the mould filling process showing the deformation of an initially square fluid element at successive positions of the advancing melt front.

(After Folkes and Russell, 1980, Polymer, 21, 1252-1258)

Experimental confirmation of this phenomenon has been reported by Schmidt (1974) using a very elegant technique. A mould having glass windows was attached to a capillary rheometer, which was loaded with a pre-formed rod of polymer, containing a series of coloured markers along its length. The movement of these coloured markers could be followed during injection using high speed photography. The sequence of colours in the moulding was found to be in reverse order to that in the original pre-formed rod, and is consistent with the process of simultaneous flow and solidification. A typical result obtained by Schmidt (1974) is shown in Fig. 6.12. This work was performed using both unfilled and short glass reinforced polypropylene and showed that the pattern of mould filling was not essentially different when fibres were present.

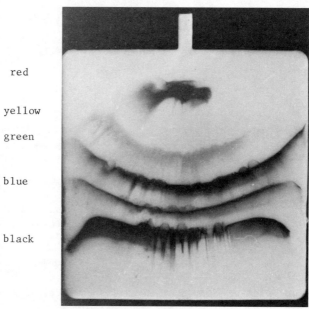

FIG. 6.12. Polybuteneterephthalate plaque moulded at constant volumetric flow rate at a melt temperature of 260°C and a mould temperature of 23°C. The red tracer entered the mould first, followed by the yellow, green, blue and black tracers respectively. (After Schmidt, 1974)

In order to interpret the origin of the fibre orientation distribution
observed in simple mouldings, it is necessary to use some of the
ideas advanced by Tadmor (1974). With reference to Fig. 6.11., it
can be seen that on entering the cavity, the fluid will initially
be subjected to compressional flow, leading to a divergence of the
stream lines away from the gate. Fluid elements/fibres will be
oriented orthogonally to the axis of the mould. Simultaneously, a
solidified layer of polymer forms on the mould surface. A flow
front establishes itself, advancing forward by the flow of the molten
polymer through a channel (often referred to as the "core" region)
defined by the boundaries of the solidified polymer (often referred
to as the "skin" layer). The original compressional flow field
in the melt at the gate is gradually replaced by a velocity profile
corresponding to the non-isothermal flow of a non-Newtonian fluid.
The detailed shape of this velocity profile depends on the rheological
characteristics of the fluid, the flow rate and the temperature of the
incoming melt, as well as the temperature of the mould cavity. In
particular, at high injection speeds, the fluid in the core will have
a very blunt velocity profile, in fact almost plug-like, so that the
fibre orientation present in the core in the final moulding will be
virtually identical to the transverse orientation developed at the
gate. At low injection speeds, where the velocity profile in the
core is less blunt, the transverse fibre orientation at the gate
will be affected by the shear flow and will result in the development
of a preferred fibre alignment along the flow direction. Further
towards the flow front, the velocity field becomes considerably
more involved. In a Lagrangian frame of reference (i.e. one in
which the observer moves with the same velocity as the advancing
flow front) the motion of the fluid/fibres is similar to that
of a "fountain", with fluid elements declerating as they approach
the flow front from the core and acquiring a radial component of
velocity as they move towards the wall. In a laboratory frame of
reference, the actual change in orientation of a fibre during
this process is shown in Fig. 6.11. The elongational flow field at
the flow front is therefore very effective in aligning the fibres
along the axis of the moulding in the "skin" layer. For the

particular case of this simple moulding and using the normal high injection speeds employed by commercial moulders, it appears that the moulding will consist of transversely aligned fibres in the core and a preferred alignment of fibres along the axis of flow in the skin layers. This is consistent with the observed fibre orientation distribution shown in Fig. 5.16.

Although the above discussion refers to a very idealised moulding, which is far removed from the real complex mouldings encountered in practice, it demonstrates how a knowledge of the fundamental rheological properties of fibre reinforced thermoplastics can provide a reasonable interpretation of the fibre orientation distribution in the final moulding. Essentially similar variations in fibre orientation through the thickness of complex mouldings are observed in those parts of the moulding that most closely resemble the geometry discussed above, providing similar materials are being compared. Simple changes in overall geometry e.g. going from a rectangular bar to an ASTM test specimen do not affect the fibre orientation distribution in any significant way even though during injection, the melt passes through a region of convergent flow before entering the gauge length of the specimen.

6.2.3. Flow singularities in complex mouldings

A sufficient body of experience has been accumulated generally to enable reasonably sensible predictions to be made about the fibre orientation distribution in flat sections of moulded components. However, in a practical moulding there may be a number of more complex flow situations existing and which more often than not limit the performance of the moulding. These rather more localised flows may arise from the presence of ribs, channels, corners and inserts. Reference has already been made to the work of Bright and Darlington (1980) and diagrams of the fibre orientation distributions at particular points of a complex box moulding have been shown in Fig. 5.20. Lockett (1980) has considered a number of flow geometries and these are shown in Figs. 6.13 - 6.16. Included in the diagrams are letters depicting the local type of flow which is expected to occur - S shear flow, C convergent flow and D divergent flow. The diagrams do not specifically include the flow that occurs early in

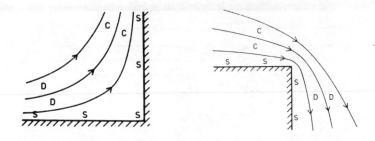

FIG. 6.13. Flow of a polymer melt around corners. In this and
successive diagrams in this series, the symbols C, D and S refer
to converging, diverging and shear flow conditions.
(After Lockett, 1980)

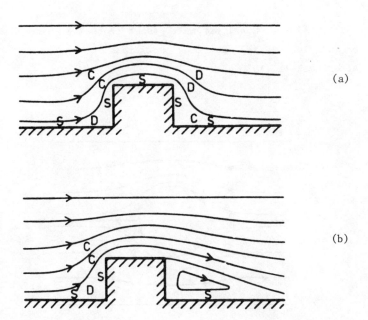

FIG. 6.14. Flow of a polymer melt past an obstacle in a mould surface
for (a) slow and (b) fast flow conditions.
(After Lockett, 1980)

138

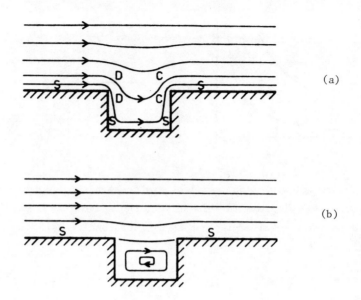

(a)

(b)

FIG. 6.15. Flow of a polymer melt past a channel in a mould surface
for (a) slow and (b) fast flow conditions.
(After Lockett, 1980).

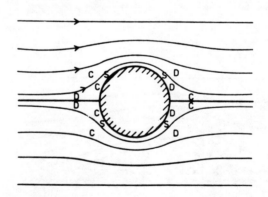

FIG. 6.16. Flow of a polymer melt past a cylindrical obstacle.
In high speed flow, circulation behind the obstacle may occur.
(After Lockett, 1980).

the process of moulding filling when simultaneous solidification takes place. Whether it is realistic to ignore this, when comparing Lockett's predictions with observed fibre orientation in moulded components, is not known at this stage in the development of the subject. Taking the particular case of the moulded-in insert shown in Fig. 6.16, then ahead of the obstacle there is diverging flow which will lead to the fibres orienting orthogonally to the streamline and in the plane of the diagram. The fibres will then align circumferentially around the insert until they reach the far side. The combination of divergent and shear flow causes the fibres to take up a random orientation, until convergent and shear flow takes over and orients the fibres along the streamline direction. It is obvious, therefore that the fibre orientation distribution is very inhomogeneous and the presence of transversely oriented fibres near the insert could seriously affect the strength of the moulding, if loads were applied along the original overall streamline direction.

140

References

Bagley, E.B. (1957). End corrections in the capillary flow of polyethylene. J.Appl.Phys., 28, 624-627.

Bell, J.P. (1969). Flow orientation of short fibre composites. J. Comp. Mat., 3, 244-253.

Bernhardt, E.C. (1959). Processing of Thermoplastic Materials. Rheinhold., New York.

Bright, P.F. and Darlington, M.W. (Nov 1980). The structure and properties of glass fibre filled polypropylene injection mouldings. Plastics and Rubber Institute Conference "Moulding of Polyolefins", London.

Chan, Y., White, J.L. and Oyanagi, Y. (1978). A fundamental study of the rheological properties of glass-fiber-reinforced polyethylene and polystyrene melts. J. Rheol., 22, 507-524.

Cole, E.A., Cogswell, F.N., Huxtable, J. and Turner, S. (1979). Fiber-foam: A rheological phenomenon and a novel product. Polym. Eng. Sci., 19, 12-17.

Charrier, J.M. and Rieger, J.M. (1974). Flow of short glass fibre-filled polymer melts. Fibre Sci. Tech., 7, 161 - 172.

Crowson, R.J., Folkes, M.J. and Bright, P.F. (1980). Rheology of short glass fibre-reinforced thermoplastics and its application to injection molding. I Fiber motion and viscosity measurement. Polym, Eng. Sci., 20, 925-933.

Crowson, R.J. and Folkes, M.J. (1980). Rheology of short glass fiber-reinforced thermoplastics and its application to injection molding. II The effect of material parameters. Poly.Eng.Sci., 20, 934-940.

Einstein, A. (1906). Ann. de Phys., 19, 289. English translation in "Investigations on the Theory of Brownian Motion", Dover, New York (1956).

Goettler, L.A. (April 1970). Controlling flow orientation in molding of short-fiber compounds. Modern Plastics, 140 - 146.

Goldsmith, H.L. and Mason, S.G. (1967). The microrheology of
dispersions. In "Rheology: Theory and Applications", Volume 4,
ed. F.R. Eirich, Academic Press, New York.

Jeffery, G.B. (1922). Motion of ellipsoidal particles immersed in
a viscous fluid. Proc. Roy. Soc., 102, 161-179.

Lee, W.K. and George, H.M. (1978). Flow visualization of fiber
suspensions. Polym. Eng. Sci., 18, 146-156.

Lockett, F.J. (Sept/Dec 1980). Prediction of fibre orientation in
moulded components. Plastics and Rubber: Processing, 85-94.

Mewis, J. and Metzner, A.B. (1974). The rheological properties of
suspensions of fibres in Newtonian fluids subjected to extensional
deformations. J. Fluid Mech., 62, 593-600.

Modlen, G.F. (1969). Re-orientation of fibres during mechanical
working. J.Mat.Sci, 4, 283-289.

Murty, K.N. and Modlen, G.F. (1977). Experimental characterization
of the alignment of short fibers during flow. Polym.Eng.Sci., 17,
848-853.

Newman, S. and Trementozzi, Q.A. (1965). Barus effect in filled
polymer melts. J. Appl. Polym. Sci., 9, 3071-3089.

Nicodemo, L., Nicolais, L. and Acierno, D. (1973). Orientation of
short fibers in polymeric materials resulting from convergent flow.
Ing. Chem.Ital., 9, 113-116.

Oda, K., White, J.L. and Clark, E.S. (1976). Jetting phenomena in
injection mold filling. Polym. Eng. Sci., 16, 585-592.

Owen, M.J. and Whybrew, K. (Dec. 1976). Fibre orientation and
mechanical properties in polyester dough moulding compounds (DMC)
Plastics and Rubber, 1, 231-238.

Parratt, N.J. (1972). Fibre-Reinforced Materials Technology. Van
Nostrand Reinhold, London.

Roberts, K.D. (1973). M.Sc. Thesis, Washington University, St Louis,
Missouri.

Schmidt, L.R. (1974). A special mold and tracer technique for studying shear and extensional flows in a mold cavity during injection molding. Polym. Eng. Sci., 14, 797-800.

Segre, G and Silberberg, A. (1962). Behaviour of macroscopic rigid spheres in Poiseuille flow. J. Fluid Mech., 14, 115-135 and 136-157.

Tadmor, Z. (1974). Molecular orientation in injection molding. J. Appl. Polym. Sci., 18, 1753-1772.

Thomas, D.P. and Hagan, R.S. (1966). Flow properties of fiber-glass reinforced thermoplastics. SPI (Reinf. Plast. Div) 21st Annual Conf. Section 3-C, Chicago.

Van Wazer, J.R., Lyons, J.W., Kim, K.Y. and Colwell, R.E. (1963). Viscosity and Flow Measurement. Interscience, New York.

Wu, S. (1979). Order-disorder transitions in the extrusion of fiber - filled poly(ethylene terephthalate) and blends. Polym. Eng. Sci., 19 638-650.

CHAPTER 7
Design Aspects

7.1 INTRODUCTION

In this chapter we will be concerned with a number of aspects
relating to the mechanical properties of short fibre reinforced
thermoplastics. Some of the mechanical properties of samples, having
a simple and well-defined fibre orientation have already been
discussed in Chapter 3. In practice, however, moulded components
will exhibit a complex microstructure; the fibre orientation
distribution will be inhomogeneous i.e. it will vary from one point
in the moulding to another, as may the fibre volume fraction. The
designer would like to be able to predict the deformation of
components, using standard design data for the material. Even if
such data existed though, it would be of questionable value, since
the stress analysis of inhomogeneous composite materials poses a
very formidable problem. In some cases where the geometry of the
moulding is fairly simple, predictions of the stiffness anisotropy
may be made using some of the techniques described in Chapter 2.
When the moulding has a complex shape, this approach is protracted
and certainly unecomonic. In this case, some alternative, and less
mathematically rigorous method of compiling and utilizing design
information is required. This somewhat more radical approach to
design in short fibre reinforced thermoplastics will be discussed in
this chapter, together with the more traditional predictive method.
In addition, we will consider the overall process of optimizing the
component properties based on our knowledge of composite mechanics.

7.2. PRACTICAL ASSESSMENT OF STIFFNESS ANISOTROPY

The usual measure of stiffness that is used is either that obtained
from a normal tensile stress-strain curve or from creep measurements,
in which case the appropriate quantity is the creep modulus, obtained
from an isochronous test. The flexural stiffness, as measured using
3 or 4 point bending, is also widely used but may be markedly
different from that obtained under simple tension, due to the
distinct skin-core microstructure observed in short fibre reinforced
thermoplastic mouldings. Finally, ultrasonic wave methods may be
used to monitor the reproducibility and anisotropy of components.

7.2.1. Simple tension tests

The traditional method of assessing mechanical anisotropy is to
cut specimens from various points in a moulding and then measure
their elastic moduli. The limitation with this approach is that
such measurements do not provide "fundamental data", since the
anisotropy is not regular and varies from point to point, both in
the plane of the test specimen and through its thickness. This is
true of even the simplest injection moulded component; each one
displays properties that are strongly influenced by the mould geometry
and the processing conditions used during the moulding process.
Nevertheless, as far as the assessment of design data is concerned,
the long established practice has been to mould dumbbell shape
tensile test bars. The bar is single end gated and the flow field
during moulding gives rise to an axial alignment of the fibres.
Darlington et al (1977) have shown that this alignment is not perfect,
as has often been assumed, and furthermore the orientation
distribution is dependent on the particular material. A contact
microradiograph and the corresponding fibre orientation distribution
for a glass fibre reinforced polypropylene test bar is shown in
Fig. 7.1. The fibre alignment in the bar means that the measured
stiffness will provide an "upper bound" to the stiffness of a more
generally oriented component, and of course will grossly overestimate the transverse properties of an oriented component. The
fact that the fibres have a length distribution and are not perfectly
aligned in the test bar means that its stiffness is significantly

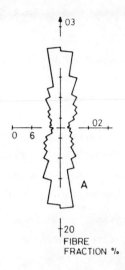

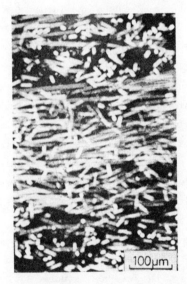

FIG. 7.1. Glass fibre orientation distribution in an ASTM testbar,
together with a corresponding microradiograph of a section cut
perpendicular to the bar axis showing that the fibres are not
fully aligned.
(After Darlington et al, 1977 and Darlington and McGinley, 1975,
J.Mat.Sci., 10, 906-909).

less than that expected for aligned continuous fibres. Darlington et al (1977) have made a thorough study of the validity of the various theories, discussed in Chapter 2, for predicting the stiffness of moulded test bars. Table 10, taken from their work, shows the results for glass reinforced polypropylene (GFPP), nylon 66 (GFPA66) and polyethylene terephthalate (GFPETP). Although widely accepted as a standard test specimen, it is clear from these data that the stiffness corresponds to a rather complex fibre orientation distribution. Furthermore, the stiffness will be dependent on the thickness of the test bar, since the proportion of the relatively highly aligned skin layers will change.

It has been proposed that tensile specimens cut with their axes at 90° to the principal flow direction in an edge gated disc should be used for measurements of the "lower bound" stiffness. A flash gated square plaque could be used to provide specimens having a range of fibre orientations - Dunn and Turner (1974). Unfortunately, it has been found in the case of injection moulded discs that the 90° specimens (see Fig 7.2) do not necessarily reveal the lowest stiffness. Thus Darlington et al (1976, 1977) have shown that for 3mm thick discs the stiffness is a maximum along the axis of flow for glass fibre reinforced nylon 66, but a minimum for glass fibre reinforced polypropylene. The situation is even more depressing than this, since if a 6mm thick disc is tested, the result for the glass fibre reinforced nylon 66 is reversed, again due to the change in relative magnitudes of the effects due to the skin and core regions in the moulding. The anisotropy of the stiffness in glass fibre reinforced nylon 66 is shown in Fig. 7.3. Although by aligned continuous fibre standards the anisotropy is modest, it is reasonably typical of the value expected for samples removed from more complex shaped components, e.g. the seed box, studied by Dunn and Turner (1974) and the display stand moulding, studied by Bright and Darlington (1980). The problem is that for the purpose of economic design in fibre reinforced thermoplastics, the levels of anisotropy shown above cannot be ignored. This contrast with the case of unfilled thermoplastics where, although some molecular orientation can arise in injection moulded components, design procedures based

MATERIAL	EXPERIMENTAL VALUES		CALCULATED VALUES — 100 sec CREEP MODULUS FOR ASTM TENSILE BARS (GNm^{-2})							
	ASTM BAR	MATRIX	ALIGNED CONTINUOUS FIBRES	FULL LENGTH DISTR.	VOLUME AVERAGE FIBRE LENGTH	MEAN FIBRE LENGTH	FIBRE ORIENTATION DISTR.	MEAN FIBRE LENGTH AND FULL FIBRE ORIENTATION DISTR.		FULL FIBRE LENGTH AND ORIENTATION DISTR.
				COX	COX	COX	COX-KRENCHEL	COX-KRENCHEL	HALPIN	COX-KRENCHEL
POLYPROPYLENE 26% wt GLASS FIBRES (23°C)	5.40	1.52	9.61	7.56	8.31	7.63	6.61	5.35	5.17	5.30
NYLON 66 32% wt GLASS FIBRES (23°C)	9.66	3.13	15.99	11.43	12.17	11.09	11.75	8.39	8.48	8.63
POLYETHYLENE TEREPHTHALATE 17% wt GLASS FIBRES (65°C)	4.59	1.91	9.30	7.12	7.66	7.14	6.42	5.08	-	5.07

TABLE 10. Experimental and theoretical tensile creep modulus values for ASTM tensile bars of short glass fibre reinforced thermoplastics.

(Data extracted from Darlington et al, 1977)

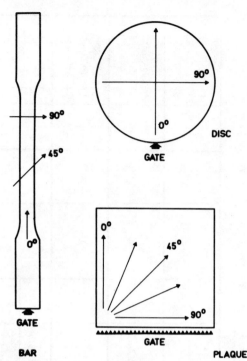

FIG. 7.2.   Definition of 0° and 90° specimens cut from injection
moulded test samples.

(After Dunn and Turner, 1974).

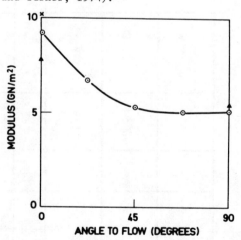

FIG. 7.3.   Anisotropy of tensile modulus in mouldings of glass fibre
reinforced nylon 66; **x** ASTM bar; ⊙ flash-gated square plaque
150 x 150mm; **△** edge gated disc 150mm diameter.
(After Dunn and Turner, 1974).

on the assumption of isotropic behaviour are reasonably valid.
Whatever their shortcomings, it seems likely that the majority of
design data for reinforced thermoplastics will continue to be based
on the "standard" specimens i.e. the ASTM bar, disc and square
plaque. An alternative testing strategy, arising from work carried
out in the laboratories at ICI Petrochemicals and Plastics Division,
will be discussed briefly in section 7.5.

7.2.2 Flexural stiffness measurements

In practice, short fibre reinforced thermoplastics will be used
under conditions of flexure as well as simple tension. In an
unfilled material, flexural modulus measurements made using 3 point
bend tests should in principle yield a value similar to the
modulus obtained in simple tension. Indeed, flexural measurements
are frequently used in preference to simple tension because of their
comparative simplicity. The state of stress throughout the specimen
will be quite different though under these two loading conditions.
In simple tension the stress is uniform across the cross-section,
whereas in flexure the stress will increase from zero on the neutral
axis to a maximum on the top and bottom surfaces of the specimen.
When the material is inhomogeneous and anisotropic, it is very
likely that the modulus measured in simple tension and flexure
will differ significantly. Such a situation could arise in mouldings
of short fibre reinforced thermoplastics, due to differences in the
fibre orientation distribution in the skin and core regions. Smith
et al (1978) have reported measurements of the flexural modulus of
samples cut from edge gated discs. In the case of a 6mm thick disc
of short fibre reinforced polypropylene, it was found that the
tensile modulus measured on samples cut at 90° to the major flow
direction was greater than that for samples cut at 0°. Flexural
modulus measurements revealed a reverse pattern of behaviour. An
extract of some of their data is given in Table 11. These results
may be interpreted using the contact microradiographs shown in
Fig 7.4., which illustrate the variation of fibre orientation through
the thickness of the disc. A well developed skin-core structure
exists. Smith et al (1978) have shown that the surface layer fibres
are oriented nearly randomly in the plane, while the core fibres are

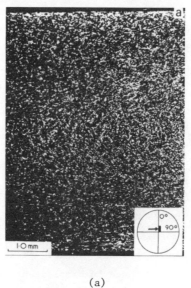

(a)

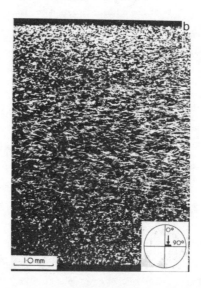

(b)

FIG. 7.4. Contact microradiographs of through thickness sections in
the 0° (a) and 90° (b) directions from a glass fibre reinforced
polypropylene injection moulded disc. For definition of 0° and 90°
specimens see Fig. 7.2.
(After Darlington, Gladwell and Smith, 1977, Polymer, 18, 1269-1274)

MODE OF TESTING	EXPERIMENTAL STIFFNESS GNm^{-2}			CALCULATED STIFFNESS (COX-KRENCHEL) GNm^{-2}		
	E_{90}	E_0	E_{90}/E_0	E_{90}	E_0	E_{90}/E_0
TENSION	4.41	3.09	1.43	5.26	3.29	1.60
FLEXURE	3.72	4.10	0.91	4.36	3.93	1.11

TABLE 11. Stiffness and anisotropy ratio data for 6mm thick edge-
gated injection moulded discs of short glass fibre reinforced
polypropylene. Simple tension results were obtained on samples cut
either parallel or perpendicular to major flow direction in the disc.
(Data extracted from Smith et al, 1978).

highly oriented in the 90° direction. As flexure is largely
influenced by the properties close to the moulding skin, while
tension reflects a through thickness average property, little
anisotropy in flexure but large anisotropy in tension is expected for
this disc, as the experimental stiffnesses confirm.

Overall, the flexure and tension comparisons demonstrate that in no
way can a knowledge of the mechanical anisotropy in one deformation
mode be used to predict anisotropy in another deformation mode, for
the same moulding. Furthermore, the observed anisotropy will again
be a function of the moulding thickness. In recognition of the
fact that for sheet material, bending is the commonest mode of
deformation that arises in service , Stephenson et al (1979, 1980)
have proposed the flexural testing of complete square plaques or discs,
instead of small samples cut from them. The advantages of this being
that little or no specimen preparation is required and because a
large specimen is examined, any problems associated with inhomogeneity
are reduced. The penalty involved in this approach is that it is
sometimes difficult to extract a flexural stiffness value from the
measurements, due to the complicated testing conditions. The square
plaque test method is a modification of the conventional three-point
bend test, in which the specimen width is of the same order as the

span. The apparatus used for these measurements consists of a variable-span flexure jig, as shown in Fig. 7.5, which is used in conjunction with a normal tensile testing machine. The load versus flexural deformation of the plate is recorded. The plate can be flexed in one position, rotated through 90° and flexed again, to provide a measure of the effective anisotropy for these two directions. For small loads, the flexural stiffness E_{flex} is given by:-

$$E_{flex} = \frac{L^3}{4bd^3} \frac{W}{\delta}$$

where L = span

b = specimen width

d = specimen thickness

W/δ = slope of the load versus cross-head movement trace.

The results obtained will be related to the gating arrangements used during the moulding of the plaque, since these will influence the pattern of fibre orientation. Results obtained for glass fibre reinforced polypropylene plaques, which have been moulded using either a single or double feed gate are shown in Table 12. This approach provides a valuable link between the processing conditions used to mould components and their properties under in-service conditions.

This technique is particularly attractive if disc specimens are used. In this case the full variation of flexural stiffness in the plane of the moulding can be assessed, since the geometry of the test remains constant. Stephenson et al (1979, 1980) have used either moulded discs or have cut them from square plaques. The gating methods employed are depicted in Fig. 7.6. Measurements were then undertaken of the flexural stiffness or load versus deflection for different directions of bending. The results are shown in Fig. 7.7 and confirm that the measurements are capable of discriminating between the various gating arrangements.

7.2.3. Ultrasonic measurements

In section 5.1.4. , the use of ultrasonic measurements as a means of assessing fibre orientation was briefly discussed. More generally though, the technique is widely used in the non-destructive testing

FIG. 7.5. Variable-span three point flexure jig for the rapid
assessment of mechanical anisotropy in injection moulded plaques.
(After Stephenson et al, 1980).

SPECIMEN	EFFECTIVE PLATE FLEXURAL STIFFNESS E_{flex} (GNm^{-2})		ANISOTROPY RATIO E_{flex} (0°)/E_{flex} (90°)
	E_{flex} (0°)	E_{flex} (90°)	
3.1mm THICK FLASH-GATED PLAQUE, DOUBLE-FEED	6.3	3:1	2.0
3.1mm THICK FLASH-GATED PLAQUE, ASSYMETRICAL SINGLE-FEED	5.8	4.1	1.4

TABLE 12. Effective plate flexural stiffness and anisotropy ratio
for short glass fibre reinforced polypropylene.
(After Stephenson et al, 1979).

154

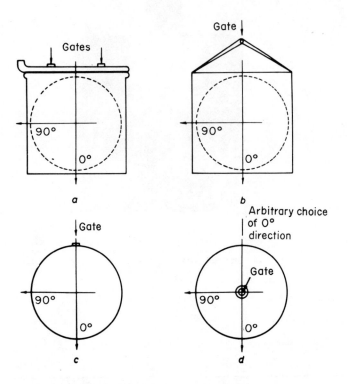

FIG. 7.6. Mould geometries and definition of 0° and 90° specimens;
(a) double-feed plaque (b) "coathanger" feed plaque (c) edge-
gated disc (d) centre-gated disc.
(After Stephenson et al, 1980).

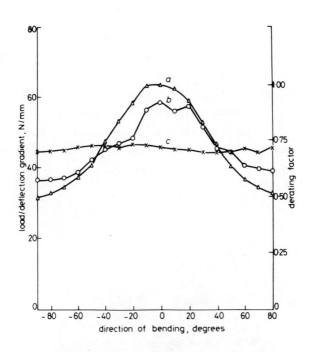

FIG. 7.7. Comparison of the anisotropy of different mouldings using a disc flexure test. Polypropylene containing 25% wt fraction of short glass fibres (a) double feed (b) "coathanger" feed (c) centre feed.
(After Stephenson et al, 1980).

field for examining the itegrity of engineering components. Properly applied therefore, the method will enable both the elastic modulus and information concerning the density (and sometimes location) of inhomogeneities, e.g. voids, to be obtained. It should be noted, however, that the elastic modulus measured at ultrasonic frequencies ~5MHz is not the same as the modulus measured in normal static tests, due to the viscoelastic nature of the matrix.

To transmit an ultrasonic wave into a material requires that the vibrations in a source transducer be transmitted into a grease or liquid to produce a longitudinal wave, which in turn is transmitted into the material. The use of such a coupling medium ensures reproducible interface conditions. In general, measurements will be facilitated if the component under study can be immersed in a bath of water, together with the source and receiver transducers. A practical arrangement is shown in Fig. 7.8., taken from the work of Thomas and Meyer (1974, 1976). This equipment has been used to

FIG. 7.8. Constant temperature water tank and rig for the ultrasonic measurement of plastics components.

(After Thomas and Meyer, 1976).

study the stiffness anisotropy in various components, especially those arising from the automotive industry e.g. gears, mounting brackets etc and which have been moulded in reinforced thermosets and thermoplastics. Measurements made on simple shapes e.g. flat discs are particularly interesting, since they provide further support to the findings discussed in the two previous sections. Accordingly, Thomas and Meyer (1974,1976) examined edge gated discs, moulded in polytetramethyleneterephthalate containing 30% by weight of short glass fibres. Visual examination of the surfaces of the discs suggested that the material fans out from the injection point during moulding and alignment of the fibres in the directions shown in Fig. 7.9.was anticipated. It could therefore be expected that for the diametral plane at 45° to the main axis of flow (as indicated in this diagram), the elastic modulus should be higher to the left of the disc and lower to the right, where fibres are expected to lie roughly parallel and normal, respectively, to the diameter. This direction is not accessible directly to the non-destructive measurement of V_L^2 (α stiffness), but measurements through the thickness in the plane shown could be made to within 30° of the disc surface. The results obtained, together with those corresponding to the case where the direction of the wave is normal to the plane of the disc, are shown in Fig. 7.10. The latter show little variation across the disc but the 30° values extrapolated into the plane of the disc imply a higher stiffness nearer the injection point with the lower value furthest from this point. This contradiction between expectation and results is again a reflection of the fact that the fibre orientation distribution varies with depth in the moulding, as confirmed by contact microradiography. The ultrasonic technique gives an average result for the volume examined. In general, therefore, ultrasonics can provide a ready means of making non-destructive quality control tests at selected points in an article and both defects and anisotropy can be examined.

7.3. LONG TERM CREEP BEHAVIOUR

Stiffness data of the type discussed above are not necessarily meaningful in relation to those applications demanding a load-bearing

158

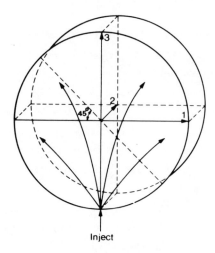

FIG. 7.9.   Schematic diagram showing the directions of texture
markings observed on the surface of glass fibre reinforced
polytetramethyleneterephthalate edge-gated discs.
(After Thomas and Meyer, 1974).

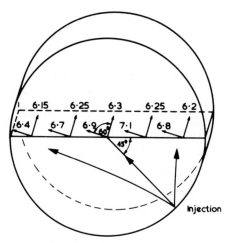

FIG. 7.10.   Ultrasonic measurements of the square of the longitudinal
wave velocity in a glass fibre reinforced polytetramethyleneterephthalate edge-gated disc.   Measurements correspond to a
diametral plane, which is at an angle of 45° to the major flow
direction.
(After Thomas and Meyer, 1974).

capability over extended periods. In such cases, the stiffness has to be replaced by a family of creep curves i.e. plots of strain versus time, for different applied stresses. The additional experimental effort can be considerable especially if the material is anisotropic. Darlington and Saunders (1970) have shown that the creep properties of specimens cut with their axes at different angles from a highly anisotropic sheet of unreinforced polyethylene varies widely. In this case, the presence of a high degree of molecular orientation radically changes the intrinsic viscoelastic response of the material. However, in a short fibre reinforced thermoplastic, where the mechanical anisotropy mainly arises from the presence of fibre alignment, it transpires that the time dependence of the creep strain is almost independent of the direction of loading, with respect to the anisotropy axis of the component. Of course the <u>magnitude</u> of the creep strain will depend on the loading direction. This is shown clearly in Fig. 7.11., taken from the work of Dunn and Turner (1974). Here the creep

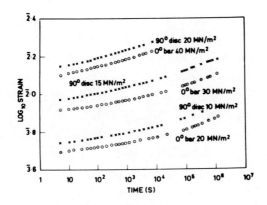

FIG. 7.11. Tensile creep curves for glass fibre reinforced
   polytetramethyleneterephthalate at 140°.
   (After Dunn and Turner, 1974).

curves are plotted as log strain versus log time, in which case the
creep curves corresponding to the "bounds" specimens appear as a
shift along the strain axis. This is an important result, since
it suggests that a considerable saving of effort and hence cost may
be made by utilizing the similarity in shape of the creep curves.
Thus a single set of creep curves for say a 0° bar specimen may be
used, together with isochronous data for other loading directions,
to generate the creep curves corresponding to any angle between the
loading direction and anisotropy axis of the component. Deviations
from this pattern of behaviour can occur, but are probably not too
serious in most cases.

The **similarity** in time dependence of the creep strain for upper
and lower bounds specimens appears to extend to the case where creep
is measured in three point bending, rather than in simple tension.
The apparatus shown in Fig. 7.5. can be used to evaluate the creep
response of wide plates, where the stress will be biaxial. So again
there is the possibility of interpolating creep curves for specimens,
having different fibre orientation distributions, using short duration
experiments.

Since by its very nature, creep testing is a protracted and costly
exercise, it would be highly desirable if the creep response of a
short fibre reinforced thermoplastic could be confidently predicted,
using the matrix creep data together with the measured fibre length
and orientation distributions. This procedure would require an
extension of the approach discussed in section 7.2.1. for predicting
the isochronous creep modulus. The validity of the theories discussed
in Chapter 2, for predicting composite properties, needs careful
examination especially when test specimens are subjected to finite
strains - Christie and Darlington (1980).

7.4. IMPACT BEHAVIOUR

The interpretation of impact energy data as measured using pendulum
methods, based on Izod or Charpy, is the centre of continuing debate
even for unreinforced polymers. When short fibre reinforced thermoplastics are tested under impact conditions, the variation of impact
energy versus fibre content is confused, in that for some polymers it

increases with addition of fibres and decreases in others. As an
example of the complexity of the situation, Figs. 7.12. and 7.13.show
the unnotched and notched Izod impact strengths for a number of short
glass fibre reinforced polymers. Nylon shows an increase in impact
strength with fibre content, for both unnotched and notched specimens,
but the behaviour of reinforced polypropylene is very dependent on
the notching conditions. Furthermore, the results can depend on
whether the notch is "moulded-in" during the injection moulding of
the specimen or cut in a separate operation. Some of the irregularity
in the impact data probably arises from variations in fibre length
distributions during the moulding of the various specimens. For
example, the addition of carbon fibres to nylon 66 increases the
overall impact strength - Theberge and Robinson (1974). Part of this
increase in energy absorbing capability may arise from the presence
of a large proportion of short fibres, which can toughen the composite
by one of the mechanisms debated in Chapter 2. Irrespective of the
origin of the effects, data obtained from Charpy or Izod tests are
really only of value for general comparative purposes.

This method of assessing the fracture toughness of reinforced polymers
has been criticised in that it does not really examine the material
as it exists in the final component, where processing and geometrical
features can lead to areas of anisotropy and weakness, which are not
tested in simple specimens. For example, the presence of fibre
orientation can result in a marked anisotropy of the impact strength
and hence contributes to the broad scatter of impact data. This
particular point has been examined in detail by McNally (1977), who
examined the angular dependence of the notched Izod impact strength
in oriented plaques of short glass fibre reinforced polybutyleneterephthalate - see Fig. 7.14. It can be seen that the impact
strength changes by a factor of about 2.5 as the direction of impact
with respect to the overall fibre direction moves through 90°. A
somewhat similar anisotropy was observed by Thomas and Meyer (1976),
who removed miniature Izod impact specimens from an injection moulded
ASTM bar of glass fibre reinforced polytetramethyleneterephthalate.
From a practical point of view these tests, although providing a
basis for a detailed understanding of impact behaviour, do not mimic
the kind of in-service impact conditions experienced in real

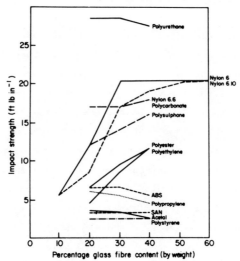

FIG. 7.12. Unnotched Izod impact strength versus glass fibre content
for a range of thermoplastics.

(After Titow and Lanham, 1975, Reinforced Thermoplastics,
Applied Science Publishers, London).

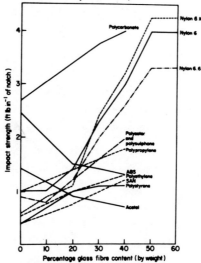

FIG. 7.13. Notched Izod impact strength versus glass fibre content
for a range of thermoplastics.

(After Titow and Lanham, 1975, Reinforced Thermoplastics, Applied
Science Publishers, London).

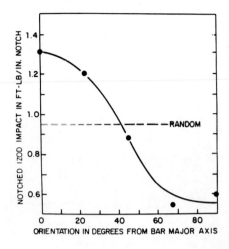

FIG. 7.14. Anisotropy of the notched Izod impact strength in
aligned plaques of short glass fibre reinforced polybutylene
terephthalate.
(After McNally, 1977).

components. For many years now, the falling weight impact test
has filled this gap, but requires a large number of specimens and
is tedious to apply when carried out properly. It does have two
obvious advantages though. One is that the test specimen can be
the complete moulding or part of it. In addition, the stress system
generated by the falling tup, when striking the specimen, is biaxial
i.e. the direction of failure is not imposed (as in a notched impact
specimen) and a crack can propagate in the weakest direction.
However, in all types of conventional impact testing, only an energy
absorbed by the material during failure is recorded. Significantly
more information concerning the fracture mechanism can be obtained
by recording the stress-time curve, using an instrumented tup. This
has been applied to the falling weight test by Dunn and Williams
(1980). Fig. 7.15. shows an overall view of the equipment, much of
which consists of the essential electronics needed to record the
stresses generated during the short duration of impact. Typical
experimental traces obtained from glass fibre reinforced polypropylene

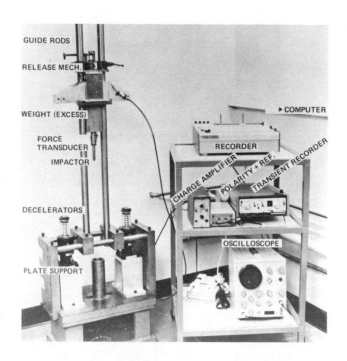

FIG. 7.15. Overall view of the instrumented falling weight impact
equipment.

(By courtesy of Dr S. Turner, ICI Plastics Division).

and nylon 66 are shown in Fig. 7.16. This shows rather clearly how

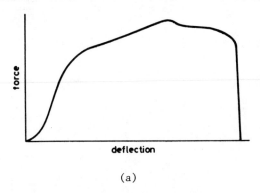

(a)

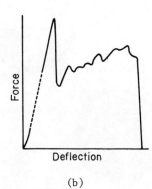

(b)

FIG. 7.16.   Force – deflection curves obtained at high rates of
loading from disc specimens of (a) glass fibre reinforced
polypropylene (b) glass fibre reinforced nylon.
(After Dunn and Williams, 1980).

the two materials could have very similar impact energies and yet
from the force deflection data the fracture mechanisms can differ
significantly. The potential of the technique for distinguishing
failure mechanisms is shown in Fig. 7.17. Although these results
correspond to an unreinforced plate, it is clear that the change from
brittle to ductile type failure, as the temperature is increased,
could be anticipated for a reinforced thermoplastic as well. This
general trend towards instrumented impact testing is providing much
new and unexpected data, which would not have been possible using

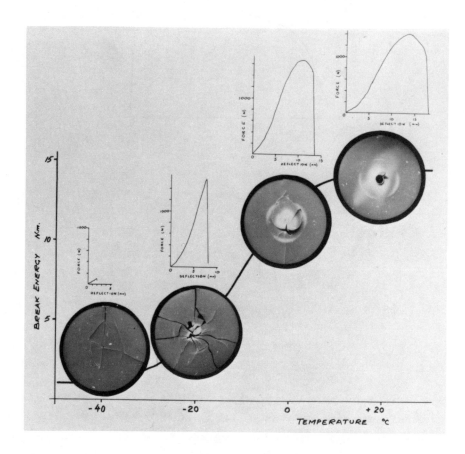

FIG. 7.17. The brittle-ductile transition of injection moulded
polypropylene discs with the corresponding load-deflection curves
as obtained in the instrumented falling weight impact test.
(By courtesy of Dr S. Turner, ICI Plastics Division).

traditional test methods.

7.5. SUB-COMPONENT APPROACH TO DESIGN

From the previous discussion, it is clear that upper and lower bounds
for the stiffness and strength of some short fibre reinforced
thermoplastics may be obtained from measurements on 0° ASTM bars and
90° disc or plaque specimens. From these data, maximum and minimum
design stresses can be derived. However, component design based on
either of these limits will be undesirable, since the upper design
stress will significantly overestimate the safe working stress in a
more generally oriented component, whereas the lower value will lead
to uneconomic over-design. In addition to the problem of allowing
for anisotropy in component design, other features occurring in the
component may demand a choice of lower design stress. One such
feature is a weld line, generated by the use of multiple gating or
when the flow path is split due to the presence of a mould insert.
Some method is clearly required whereby anisotropy and other
processing artefacts can be incorporated in the detailed design
procedure. One such approach is to provide the designer with a table
of data giving the amounts by which the nominal design stress is
reduced or "derated" as a result of the presence of these moulded in
features. It is with these ideas in mind, together with a general
dissatisfaction with standard methods of evaluating the design
properties of reinforced thermoplastics, that Stephenson et al (1979)
propose that end products be regarded as combinations of several basic
"sub-components". These could be a long thin channel, a wide channel,
a region of converging flow, a region of diverging flow, an internal
weld (butt weld and knit line) etc. Experimental data obtained for
various common mould geometries is given in Table 13, where the
anisotropy derating factor A is referred to the 0° ASTM bar value.
In this context, the derating factor is simply the ratio of design
stress for the particular geometry to the design stress for a 0°
ASTM tensile bar. The advantage of this approach is that once
corrections have been made for the mean level of stiffness for a
particular fibre concentration, the major factors affecting the
distribution of stiffness throughout any given component will be the
mould geometry and gating arrangements, and these can be taken into

Mould geometry and feeding system	Direction (Dunn-Turner nomenclature)	Thickness	Deformation mode	Anisotropy derating factor, A
1. End-gated bar or strut 0° direction	0°	3 mm	Tension Flexure	1
	0°	6 mm	Tension	0·85
	0°	12 mm	Tension	0·45
2. End-gated plaque 0°	0°	3 mm	Tension Flexure	0·9
	90°	3 mm	Tension Flexure	0·6
	0°	6 mm	Tension Flexure	0·85
	90°	6 mm	Tension Flexure	0·50
3. Centre gate radial flow radial / tangent	radial	3 mm	Tension Flexure	0·55 0·65
	tangent	3 mm	Tension Flexure	0·80 0·65
	radial	6 mm	Tension Flexure	0·5 0·6
	tangent	6 mm	Tension Flexure	0·7 0·6
4. Butt weld in bar	0°	3 mm	Tension Flexure	0·6
5. Double-feed plaque (includes side-side weld) 0° 90°	0°	3 mm	Tension Flexure	1·0
	90°	3 mm	Tension Flexure	0·5
	0°	6 mm	Flexure	0·8
	90°	6 mm	Flexure	0·5
6. Torsion Take shear modulus to be 0·2 of bar tensile value.				

Note. Welds assume good venting and suitable flow lengths.

TABLE 13. Anisotropy derating factors for polypropylene containing 25% wt fraction of short glass fibres.

(After Stephenson, May 1979, Plastics and Rubber: Materials and Applications, pp45-51).

account using the type of data given in Table 13. The information obtained from tests on sub-components will be less fundamental than that obtainable on specimens of ideal geometry, but it is expected to be of greater relevance to practical design. Furthermore, it should provide a more effective route for the optimization of tool design and processing conditions. The general scheme for the new method is compared with that for the traditional method in Fig. 7.18.

7.6. OPTIMISATION OF COMPONENT PROPERTIES

Throughout this book we have been concerned with the principles upon which stiff, strong and tough short fibre reinforced composite components may be produced. The overall optimisation of the component and its properties is the result of the following three design exercises:-

(a) Design of the basic material - choice of matrix, fibre, fibre concentration, fibre length/distribution, fibre/matrix bond strength etc.

(b) Engineering design of the component - selection and presentation of design data for the composite, utilization of design data in the prediction of both the short and long term mechanical properties of a given component and its affect on the choice of component dimensions.

(c) Process design for component fabrication - choice of fabrication method (usually injection moulding), processing conditions to be used, location of the gate(s) in the mould etc.

Usually though there are other requirements imposed on the component, such as the need for weight saving and the almost universal need that the component be produced as cheaply as possible. If for example, the component is to be subjected to mainly flexural loading, then considerable weight saving is possible by the use of a low density core. This is really the principle of the I beam, where the volume of load bearing material is reduced near the neutral axis, where the stresses are small. A low density core in a plastics component can be achieved by the use of blowing agents, which produce large volumes of gas during component fabrication and lead to a cellular structure in the core of the moulding - see Fig. 7.19.

TRADITIONAL METHOD

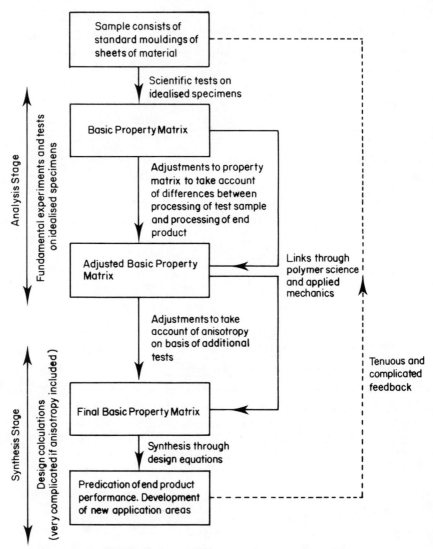

FIG. 7.18.   Schemes for data generation and design.

(After Stephenson et al, 1979)

NEW METHOD

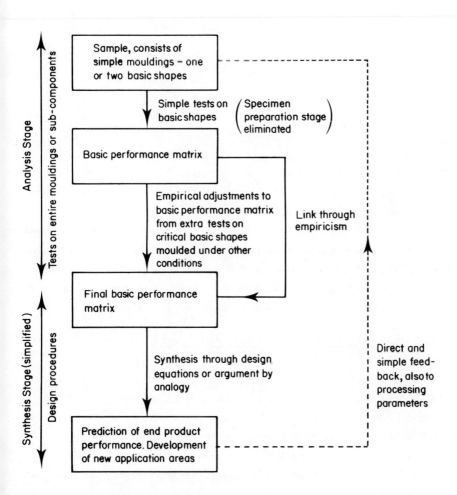

FIG. 7.18. Schemes for data generation and design.
(After Stephenson et al, 1979).

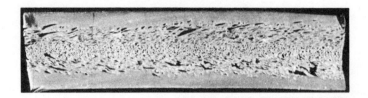

FIG. 7.19.   Through thickness section cut from a sample of structural
   foam.

This can be applied to reinforced thermoplastics, where the presence
of the short fibres leads to improved control of the cell size and
furthermore the fibres tend to orient parallel to the cell walls,
giving improved rigidity to the foam - Wilson (1971). The stiffness
of reinforced thermoplastic structural foams may be calculated using
the theoretical methods discussed in Chapter 2, providing the stiffness of the cellular core can be calculated using, for example, the
empirical relationship proposed by Moore et al (1974):-

$$\frac{\text{Stiffness of foam}}{\text{Stiffness of skin}} = \left(\frac{\text{density of foam}}{\text{density of skin}}\right)^2$$

Further improvements in the overall optimization of the component
may be possible by the use of more than one fibre species. The
potential advantages of "hybrid" composites, containing two types of
intimately mixed fibre have been discussed by Richter (1977). As
mentioned earlier in this book, the presence of a proportion of
fibres having a length $\approx \ell_c$ can significantly improve the toughness of
the composite. However, if the fibres are carbon then it is obviously
very uneconomic to use a proportion of these for toughening. It is
sensible in this case to use cheaper fibres such as glass for the
toughening role and retain the carbon for producing a stiff and
strong composite. This approach has been shown to be a viable method
of improving the properties of a short fibre composite - Folkes and
Sharp (1981).

References

Bright, P.F. and Darlington, M.W. (Nov 1980). The structure and properties of glass fibre filled polypropylene injection mouldings. Plastics and Rubber Institute Conference "Moulding of Polyolefins, London.

Christie, M.A. and Darlington, M.W. (Aug. 1980). Third International Conference on "Composite Materials", Paris. Published in "Advances in Composite Materials" ed. A.R. Bunsell et al., pp 260-275,Pergamon Press, Oxford.

Darlington, M.W., Mc.Ginley, P.L. and Smith, G.R. (1976). Structure and anisotropy of stiffness in glass fibre reinforced thermoplastics. J. Mat. Sci., 11, 877-886.

Darlington, M.W., McGinley, P.L. and Smith, G.R. (May 1977). Creep anisotropy and structure in short-fibre-reinforced thermoplastics: Part I Prediction of 100 sec creep modulus at small strains. Plastics and Rubber: Materials and Applications, 51-58.

Darlington, M.W. and Saunders, D.W. (1970). The tensile creep behaviour of cold-drawn low density polyethylene. J.Phys. D., 3, 535-549.

Dunn, C.M.R. and Turner, S. (1974). The characterization of reinforced thermoplastics for industry and engineering uses. In "Composites- Standards, Testing and Design," NPL, England, pp 113-119, IPC Science and Technology Press, Guildford.

Dunn, C.M.R. and Williams, M.J. (May 1980). Measurement of the strength of thermoplastic plates. Plastics and Rubber: Materials and Applications, 90-96.

Folkes, M.J. and Sharp, J. (1981). To be published.

Mc.Nally, D. (1977). Short fiber orientation and its effect on the properties of thermoplastic composite materials. Polym.-Plast. Technol. Eng., 3(2), 101-154.

Moore, D.R., Couzens, K.H. and Iremonger, M.J. (1974). Deformational behaviour of foamed thermoplastics. J. Cell. Plast., 10, 135-139.

Richter, H. (1977). Hybrid composite materials with oriented short fibres. Kunststoffe, 67, 739-743.

Smith, G.R., Darlington, M.W. and McCammond, D. (1978). Flexural anisotropy of glass-fibre-reinforced thermoplastics injection mouldings. J. Strain Anal., 13, 221-230.

Stephenson, R.C., Turner, S. and Whale, M. (1979). The load - bearing capability of short-fiber thermoplastics composites - a new practical system of evaluation. Polym. Eng. Sci., 19, 173-180.

Stephenson, R.C., Turner, S. and Whale, M. (Feb 1980). The assessment of flexural anisotropy and stiffness in thermoplastic - based sheet materials. Plastics and Rubber: Materials and Applications, 7-14.

Theberge, J.E. and Robinson, R. (Feb 1974). Carbon fibers add muscle to plastics. Machine Design.

Thomas, K. and Meyer, D.E. (1974). Anisotropy arising from processing of reinforced plastics. In "Composites - Standards, Testing and Design," NPL, England, pp 131-139, IPC Science and Technology Press, Guildford.

Thomas, K. and Meyer, D.E. (Sept. 1976). Ultrasonic measurement of reproducibility and anisotropy of processed polymers. Plastics and Rubber: Materials and Applications, 136-144.

Wilson, M.G. (1971). Guide to working with reinforced thermoplastic foams. SPE Journal, 27, 35-39.

INDEX